DYING TO WORK

DYING TO WORK

Death and Injury in the American Workplace

JONATHAN D. KARMEL

ILR PRESS
AN IMPRINT OF
CORNELL UNIVERSITY PRESS
ITHACA AND LONDON

First published 2017 by Cornell University Press

Printed in the United States of America

Library of Congress Cataloging-in-Publication Data

Names: Karmel, Jonathan D., 1955– author.
Title: Dying to work : death and injury in the American workplace /
    Jonathan D. Karmel.
Description: Ithaca : ILR Press, an imprint of Cornell University Press,
    2017. | Includes bibliographical references and index.
Identifiers: LCCN 2017021445 (print) | LCCN 2017023742 (ebook) |
    ISBN 9781501714382 (pdf) | ISBN 9781501714375 (ret) |
    ISBN 9781501709982 (cloth : alk. paper)
Subjects: LCSH: Industrial accidents—United States. | Industrial safety—
    United States. | Occupational mortality—United States. | Hazardous
    occupations—United States. | Industrial hygiene—United States.
Classification: LCC HD7262.5.U6 (ebook) | LCC HD7262.5.U6 K37
    2017 (print) | DDC 363.110973—dc23
LC record available at https://lccn.loc.gov/2017021445

Cornell University Press strives to use environmentally responsible
suppliers and materials to the fullest extent possible in the publishing
of its books. Such materials include vegetable-based, low-VOC inks
and acid-free papers that are recycled, totally chlorine-free, or partly
composed of nonwood fibers. For further information, visit our website
at cornellpress.cornell.edu.

To the working women and men of America

Pray for the dead and fight like hell for the living.

MOTHER JONES

CONTENTS

PREFACE

The 2016 election sent shock waves through both sides of the political divide and into all recesses of American life. Seeing the emergence of a strong antiregulatory agenda and the deconstruction of the administrative state, advocates for safe workplaces now fear an end to the regulatory progress made in the Obama administration to protect the health and safety of workers. Indeed, some of the very first actions taken by the Trump administration were to repeal common sense rules that would make workers safer and employers more accountable.

Completing this book in the midst of this unprecedented time has been frustrating and challenging. We cannot predict the future, but at the moment the prospects for continuing reform look dim. That does not negate the value of progress made to date nor the statistics that document the breadth and depth of the problem. This book stands as a document of the state of worker safety programs in place at the end of Barack Obama's administration. Unfortunately, that may be a high-water mark for comparison as administrations change and different ideologies prevail. Regardless, it is essential for readers, and indeed all citizens, to question whether American workers are better off and safer under an antiregulatory and pro-business agenda. Read the stories in this book and ask yourself: Are we doing everything possible to protect American workers from death and injury? And, if not, why not?

DYING TO WORK

Introduction

> Here was a population, low-class and mostly foreign, hanging always
> on the verge of starvation, and dependent for its opportunities of
> life upon the whim of men every bit as brutal and unscrupulous as
> the old-time slave drivers; under such circumstances immorality was
> exactly as inevitable, and as prevalent, as it was under the system
> of chattel slavery. Things that were quite unspeakable went on
> there in the packing houses all the time, and were taken for granted
> by everybody; only they did not show, as in the old slavery times,
> because there was no difference in color between master and slave.
>
> UPTON SINCLAIR, *THE JUNGLE*

At my place of employment, a busy law firm, the work can be demanding, the hours long and grinding. But it is, in large part, safe and unlikely to cause me any physical harm. For sure, lawyers, accountants, and other professionals and office workers may suddenly drop dead at their desks from aneurysms and ventricular fibrillation, or suffer some less deadly and common fate from stress or high blood pressure. But the most often reported injury to office workers is musculoskeletal, caused by sitting hunched over a keyboard for hours on end, word processing and surfing the Internet. After that, office injuries are caused by knocking objects from shelves, lifting file boxes, and bumping into open desk drawers. In other words, an office is a relatively safe place to work.[1] Workers there are unlikely to die from an explosion or electrocution. Workers in an office are unlikely to become fatally ill from exposure to some toxic substance, or from the air they breathe at their cubicle.

The same cannot be said for millions of others working in America[2] who take care of us when we are sick, keep us safe in our persons and in our homes, build and repair our roads and infrastructure, provide us with shelter, teach our children, grow and make our food, sell and serve our food, clean our streets and buildings, make our hotel beds, get us from one place to the next, assemble our cars, and keep us virtually connected to one another. These workers daily affect our lives, often face to face. Without them, our lives as we know them today would be unrecognizable. The simplest experience of buying milk, eating a steak, or lying down to sleep in an upscale hotel room was brought to us unobtrusively and seamlessly by someone who risked injury and death in the workplace. Yet most Americans give little thought to the real and lurking dangers in the workplace that their friends, neighbors, and family are exposed to. Instead, we are made to fear by politicians and the media much more remote dangers in our lives.

The risk of workplace death is much greater than dying in a plane crash, or being a victim of a terrorist attack. The odds of dying in a plane crash are 1 in 11 million.[3] The odds of being killed in a terrorist attack in the United States are 1 in 20 million.[4] Yet, since 2001, the U.S. government has spent more than $1 trillion in antiterrorism measures, excluding the wars in Iraq and Afghanistan.[5] For these improbable events, we spend considerable more time, treasure, and worry than we do about the very real and personal risk to, for example, a hotel housekeeper. For workers in America, the workplace is a dangerous House of Horrors. Some would say it is a jungle.

Upton Sinclair's *The Jungle*, published in 1906, was dedicated to "the Workingmen of America." It was intended as an exposé of the lives and working conditions of immigrant workers, with vivid passages describing meatpackers falling into rendering vats and being sold for lard. However, the immediate reaction to the book had less to do with the safety of the meatpackers than with food safety. The public hue and cry from *The Jungle* produced the passage of the Meat Inspection Act and the Pure Food and Drug Act of 1906, the predecessor to the Food and Drug Administration. Comprehensive national legislation protecting worker safety would have to wait sixty-four years.

In the United States today, we have a complex web of federal, state, and local laws and regulations that are intended to protect workers from harm at their workplaces. But do they really? This regulatory structure appears to have meaningful laws and regulations but is left toothless by underfunding, the inability to enforce the laws because of a lack of resources, penalties that don't deter, and by the deliberate underreporting of workplace deaths and injuries. The defanging of worker safety laws is an act of political negligence brought to American workers by a powerful business lobby, spearheaded by the U.S. Chamber of Commerce and its allies. Worker safety laws and regulations are demonized by lobbyists and politicians of both parties as job killers, while real flesh-and-blood workers die, are seriously injured, or are exposed to deadly carcinogens every day on the job.

As a result, workplace death and injuries occur daily, and in plain view, to our loved ones, friends, and neighbors. No longer are worker deaths and injuries hidden behind plant gates, protected by armed guards. They are occurring right in front of us in public and in seemingly safe workplaces everywhere. No longer are deaths and injuries occurring only to workers handling known dangerous equipment or chemicals. Deaths and injuries happen to workers in all occupations. Miners, construction workers, oil and gas workers, and railroad workers have always been the poster children of workplace death and injury. Their jobs are knowingly dangerous. But there is another class of workers—including a growing group of millions of service workers—whose jobs may seem benign but can be fatally dangerous. They work in grocery stores, hotels, hospitals, and other public places, and the workers there are frequently the most vulnerable in the workforce. They are often undocumented, workers of color, women, and minimum-wage workers, with little or no benefits and protections. Yet these workers are hidden in plain sight because Americans choose not to see them, or because we are numbed by our own powerlessness to effectuate any meaningful change.

In 2015, the last year with complete data at the time of this writing, the U.S. Bureau of Labor Statistics (BLS) reported that 4,836 workers were killed on the job, or 13 workers every day. The 2015 fatality rate was an increase from 2014, and the highest since 2008. Add to this that the Centers for Disease Control and Prevention report that annually 50,000 deaths are attributed to work-related illnesses—an average of 137 deaths each day.[6] Do the math. One hundred and fifty workers die each day because of their work. Compare that to forty-five, which is the total number of deaths in the United States since 9/11 that have any tangential relationship to jihad.[7]

Moreover, as our workforce grows increasingly part-time, the number of contract workers, or temporary workers, has grown as well. In 2015, 829 contingent workers died on the job, an increase from 2014, accounting for nearly 17 percent of all fatal work injuries. Other ignoble highlights from 2015 include a 13 percent increase in fatal injuries among women, and 409 workplace homicides. Workplace violence continues to be a growing problem, in 2014 causing 26,540 lost-time injuries, and women workers suffered 66 percent of lost-time injuries due to workplace violence.[8] All totaled, in 2015 more workers died in the United States from their work than in 2014. As for nonfatal work injuries, the number has remained stubbornly flat in recent years. In 2014, more than 3.8 million workers reported work-related injuries and illnesses.[9] In 2015, the incidence rate of nonfatal work injuries requiring days away from work to recuperate was only marginally down, from 107.1 cases per 10,000 full-time workers in 2014 to 104.0 cases in 2015.[10] Workplace injuries and illnesses are an enormous cost to the American economy of upward of $300 billion every year.[11]

Ten years into the new century, twenty-nine coal miners were killed in an explosion at the Massey Energy Upper Big Branch mine in West Virginia. It was the worst mine disaster in forty years. That same year, the front-page disasters kept coming. The BP/Transocean Gulf Coast oil platform exploded, hurling eleven oil platform workers to their deaths in burning waters, setting off a calamitous environmental and economic disaster.

What can we make of these events and statistics? With more than 150 million working in the United States, are these acceptable numbers? Arguably, the workplace is safer than it has ever been, so isn't that a good thing? Are work injuries just part of the cost of doing America's business? Life is risky, and work is part of life. Crossing the street has its risks. All these statements may be true, but they are beside the point. Instead, shouldn't we be asking whether we are doing everything that we can to make the workplace as safe as it can be? Can we do more? And, if we can, why have we failed?

It is one thing for me to make my way to work every morning, safely belted in an air conditioned car, later to be entombed in the protective shell of a downtown skyscraper office. It is quite another thing to get up every day and head back to your workplace in a hotel, hospital, or grocery store not knowing whether this may be the day that you get seriously injured, possibly fatally, often for only minimum wage, and with a tattered safety net of protections for you and your family.

But before we can make any change in this dynamic, we have to better understand the enormity of the problem. Let's start with the proposition that the health and safety of workers in America are matters of social justice, which, broadly defined, recognizes the humanity in all and our right to equal treatment in society and a fair allocation of its resources. All work and workers should be respected, whether the Nobel Prize winner or the day laborer. Yet, as I describe in this book, for most workers in the United States the right to a safe and healthy workplace has been made difficult to achieve over our country's history and remains so today in the twenty-first century.

I came to the acute awareness of the issue of workplace deaths and injuries embarrassingly late, even though I have spent more than thirty years as a union-side labor lawyer. Given my background and career choice, how did this happen? Well, for one, my focus broadly as a lawyer was in representing workers and unions in their organizing efforts and in collective bargaining, and on behalf of workers who had been wrongfully terminated. While these are kissing cousins to the issue of workplace safety, they are far enough removed that even for me, workplace safety was, more frequently than not, off my radar screen. Sure, I saw the headlines and knew that workers got injured and killed on the job. But those workers knew their risks, didn't they? Workplace death and injury were remote events, weren't they? What I didn't know is how much I really didn't know. I didn't really know the breadth and scope of the problem. I didn't really know that workplace safety was a matter of social justice, just as important as the right to organize for better wages and to be free from discrimination in the workplace, my focus. In the end, for me there wasn't a singular epiphany. I came to my awareness slowly, which led me to explore the issue more deeply and, eventually, to this book.

In getting there, I learned from many advocates for worker safety, some of whom are colleagues of mine in the labor movement. I learned that apart from new laws and better enforcement (just to name a couple of obvious reforms), maybe it would help if we—Americans who go to work every day—wake up from our collective slumber and demand a safer workplace. But I know that social change does not just happen. It first requires, as I learned, a real awareness of the problem, an awareness that is currently lacking for most of us and, as a result, makes any broad debate about workplace safety impossible, drowned out by the noise and money of corporate and political interests hostile to change. Numbers and statistics are important to a point, and this book is chock full of them. However, they also act to anesthetize us from meaningful discussion and action, all the while sanitizing the dangerous conditions in which millions work in America. American workers have faces, names, families, homes, and personal histories. They are more than numbers and statistics in a Bureau of Labor Statistics report. Numbers and statistics alone prevent us from connecting with workers and their families who suffer real and horrific workplace injuries and fatalities. Behind the number 4,836 are husbands and wives, sons and daughters, life partners, and a permanent wake of grief and loss. And because change cannot occur by distancing oneself, a physical immediacy or nearness to the workers and their families is required to achieve awareness.

This book is set out in three parts. First, I discuss the problems of achieving a safe workplace, including a brief history of workplace safety laws in the United States, and the corporate and political forces that stand in the way of change. In the end, I offer some ideas and reforms that may make workers safer and healthier than they are today, and other next steps. Bookended by these sections is the heart of the book, a collection of stories about real workers who were killed and injured in their workplaces.

For more than three years, I traveled the country and met with injured workers and surviving family members to listen and become proximate.[12] I sat in the living rooms and at the kitchen tables of surviving family members, and met with them at coffee shops and at union halls. I listened to wives, sons and daughters, and sisters and brothers tell me about their losses. I sat with injured workers, now physically disfigured and broken, and listened to how their injuries have unalterably changed their lives. For them, in the telling of their stories, all different in the cause of their

injuries, there was a common refrain that was a miraculous shared sentiment in the midst of all the despair. To a person, each and every injured
worker told me the same deepest wish. And it wasn't what I expected to
hear. I expected, as a lawyer, to be told that they wanted to be awarded
large sums of money for their injuries, their pain and suffering. Some reluctantly and, for the most part, unsuccessfully sought legal relief. But
money was not their wish. Instead, they all told me: "I just want to work."
In these most intimate settings, the family survivors and injured workers
all shared their stories and family pictures, and their tears, fears, anger,
and hopes. They gave me dog-eared file folders bound by rubber bands
and stuffed with newspaper clippings, coroner reports, OSHA records,
and court filings, entrusting to me these precious totems of their suffering.
We exchanged phone calls and e-mails. I became proximate. I got as near
as I could. And, finally, I began to understand.

My fondest wish for this book is that their stories will give readers a
nearness to the experience of these workers and their families. Proximity
is the beginning of awareness, and a necessary starting point of change.

Some of these workers' injuries or deaths were several years in the past
at the time I wrote down their stories, while others are closer in time to the
present. But all the stories are contemporary in the sense that preventable
deaths and injuries like those retold in this book occur every day. I wrote
this book over a three-year period. And during the writing, around fourteen thousand Americans died preventable deaths from their work. The
stories are also contemporary in the sense that the grief and suffering
never completely go away. I was put in contact with the survivors and the
injured workers from many sources, including from introductions made
by Tammy Miser and Tonya Ford at the United Support and Memorial
for Workplace Fatalities, where they do great work bringing attention to
these tragedies and great comfort to the families. In all cases, the families and injured workers entrusted me—a total stranger—with their most
intimate and painful feelings, and I hope that I have done justice to their
stories.

For many of the witnesses in this book, the American dream is elusive
and illusory. Along with the millions like them, the workers in this book
were only trying to get by, put food on the table and shelter over their
heads. The workers that you will meet, and workers like them all around

us, are on the front lines of a war of attrition. But they are not soldiers, where injury and death are known, if not acceptable, risks. These Americans want only to work and to return home safely at the end of the day. The workers in this book are not heroes, except to their loved ones. And workers like them are your loved ones, your neighbors, and your friends. They are all around us. They are you.

1

AMERICA GOES TO WORK

Those who cannot remember the past are condemned to repeat it.
GEORGE SANTAYANA, *THE LIFE OF REASON*

In order to understand where America is today in terms of worker health and safety, as well as where it needs to change, it is important to understand the arc of history in the American workplace.

America before the Civil War was largely a rural agrarian society. Americans lived in isolated communities, tenuously connected to others living within a short distance by horse-drawn wagons traveling over poor and primitive roads. These Americans, living in isolation, were self-sufficient in housing, food, clothing, and other life-sustaining essentials. The farm was the primary workplace, staffed by family members. There, almost all of one's needs could be manufactured or grown. The industrial sector, as it was until around 1870, consisted mostly of small firms and workshops that relied on artisan technology to produce goods for local consumption. In communities with a river for a power source, there were small industries, primarily sawmills and grain mills.

Although there is little reliable information on worker safety from back then, the Eden of the pre-industrialized America could be a mean and nasty place. Preindustrial workers risked injury from animals, hand tools, ladders, and water wheels. But for the most part, worker injuries were infrequent and thought to be the fault of the victim, who most often was also the "employer." This all changed with the onset of the Industrial Revolution.

The workplace that Americans found at the end of the nineteenth century was created in the cauldron of the Industrial Revolution, beginning well before the Civil War. Between the Civil War and World War I, and fueled by a historic wave of immigration, the United States rapidly intensified its transformation from a rural-based economy to an industrial powerhouse, centered in its growing and teeming cities. There is little dispute among historians and economists that the American Industrial Revolution occurred because of the embarrassment of natural resources, the emergence and development of American-style manufacturing (including the rise of the managerial firm), the growth of the railroads and lowered costs of transportation, and the education of the workforce. But none of this would have been possible without the more than thirty-three million immigrants who, from 1820 to 1920, landed on the shores of America, mostly from Europe, seeking the promises of the United States.

In 1880, almost one-half of American workers were farmers. Less than 15 percent worked in any kind of manufacturing. By 1920, only a mere forty years later, the numbers were almost dead even. In this same period, manufacturing employment, centralized in growing cities, increased from 2.5 million workers to 10 million by 1920. While a portion of this workforce was a product of rural-to-urban internal migration, mostly it was a result of the flood of external immigration. More than 14 million foreign-born workers were the human fuel that powered the American Industrial Revolution. Counting their children during this forty-year period, 23 million strong, more than one-third of the 105 million Americans in 1920 were first or second generation. In 1900, three-quarters of the population in most large cities were immigrants and their children.[1]

With this massive influx of immigrant labor into American cities, together with the power, transportation, and communication revolutions, the pieces were all falling in place for the industrial transformation of America. Electrical power replaced steam; railroads expanded and connected manufacturing output to markets all over the United States; telephone and telegraph altered the meaning of time and space. The final piece was the development of the organizational firm. Large corporations were located in urban cities, where the source of cheap labor lived. Giant corporations developed and became the prototype of what would become a corporate society. American corporations became more formalized, organized, and integrated.

In 1925, half a century after the end of the Industrial Revolution, Calvin Coolidge surveyed America, whose transformation he had witnessed firsthand, and declared: "After all, the chief business of the American people is business. They are profoundly concerned with producing, buying, selling, investing and prospering in the world. I am strongly of the opinion that the great majority of people will always find these the moving impulses of our life." Some have argued that this statement has often been unfairly used by his detractors as evidence of Coolidge's pro-business philosophy. Fair or not, it is an accurate and clear-eyed description of the United States as an industrial and economic colossus, embodying the world's richest and most powerful industrial nation. But at what price?

The labor force that arrived on steamships from ports all across the Atlantic were not necessarily lured by the promise of factory jobs in American cities. They were mostly unskilled laborers, farmer and artisans with very little, if any, factory experience. The potato famine in Ireland, and crises and privation in other parts of Europe, were the primary causes of immigration at the end of the nineteenth century. It was in this period that the unskilled immigrant laborer became the "dominant factory manufacturing labor force."[2]

Fleeing from famine and other adversity, this nascent industrial workforce, largely unskilled and uneducated, was tossed into the grinder of America's Industrial Revolution. Immigrant workers who in Europe had only used small hand tools and animal-powered plows and wagons were now operating unguarded mechanical equipment, powered by steam, and later electricity, in a high-speed factory setting.

Enormous manufacturing output and productivity were spawned in the Industrial Revolution. But so were dangerous working conditions previously unknown in the history of humankind. And this increased output correlated with increased worker injury and death. As American industrial might grew to unprecedented heights, producing material riches for its owners and creating a consumer society, the risk of dying or becoming seriously maimed in the workplace grew as well. This all occurred within a legal and regulatory climate that did not exactly encourage an employer to be concerned about the safety of his workers. As a result, American production methods were extremely productive, and extremely dangerous. Workers initially had little or no say about their safety, and legal liability for workplace injuries was usually shifted to the employee under assumption of risk or negligence theories, thereby making compensation for injuries nonexistent. Injuries were cheap, and workers were replaceable. There was simply no economic incentive for employers to create a safe workplace.

Nowhere was this correlation between increased production and dangerous working conditions as stark as in American coal mines and on its railroads, especially compared to their counterparts in Great Britain. Some of this can be explained by differences in mining methods, and the vast geography of the United States that railroads had to travel. But it is undeniable that in the decades immediately before and after the turn of the century, American workers were getting injured in these jobs at twice the rate of English workers. American mines yielded more coal per worker than British mines, but at double the injury rate.[3] On the rails, geography and low population density turned American railroads into primarily freight haulers, a far more dangerous business for workers than hauling passenger traffic. "The slaughter of railroad employees began almost as soon as the first lines were built."[4] Derailments and collisions were common and deadly, owing to the lack of signals and the poor condition of the track and rail bed. Worse, men had to work between moving train cars to couple and uncouple, and to work the brakes. At the end of the nineteenth century, railroad workers experienced an extraordinary level of risk, with a fatality rate of 3.14 per thousand, and likely much higher because of underreporting. By the new century, dubbed the Century of Progress, the slaughter continued. In 1907 alone, accidents killed 4,534 railroad workers.[5]

Mines and railroads were not the only dangerous workplaces. Garment workers, mostly Jews from Eastern Europe, were employed in sweatshops up and down the East Coast, but primarily in New York City, the home of the garment industry. By the end of the first decade of the twentieth century, more people worked in factories in Manhattan than in all the mills and plants in Massachusetts. Most of the workers there were employed in the garment industry.

Looking back, in 1791, treasury secretary Alexander Hamilton esti-mated that more than two-thirds of all clothing in America was home-made. Little changed over the next fifty years until Elias Howe developed the lockstitch sewing machine in the mid-1840s. This innovation made strong, straight seams and made possible the mass production of com-mercial clothing manufactured in a factory. Courtesy of the Civil War, de-mand for mass-produced clothing was created as hundreds of thousands of soldiers wore the same uniform manufactured and cut to standard sizes, differentiated only by blue and gray. The war experience, horrific as it was, spelled the death knell of homespun clothing. Such laboriously made clothing was eventually replaced by the convenience and quality of manufactured clothing, easily purchased off-the-rack in America's new department stores, or through the ubiquitous catalog. The next technolog-ical change occurred with the invention of the cutter's knife in the 1870s, which enabled a garment worker to cut pieces for identical garments in a few strokes.

With the technology in place, the millions of skilled and unskilled im-migrant workers arriving daily in America from Eastern Europe, Russia, and Italy provided the final piece to the rise of the manufacturing garment industry.[6] By 1900, homemade clothing was a preindustrial relic. In its place was a booming garment industry, centered in Manhattan, and de-pendent on immigrant workers, mostly women, who worked twelve-hour days, seven days a week, for a few dollars a day. Garment workers were crowded into dark and squalid tenement rooms and hallways on the Lower East Side, poorly ventilated, and with locked exits. The tenements became known as sweatshops, not so much for their deplorable conditions but because of the practice of "sweating" the workers for more work and less pay. In addition to receiving low wages for piecework, workers were charged for needles and thread and for the use of old sewing machines and the privilege to pump the sewing pedal with their feet for hours on end. At the end of their shifts, workers were lined up at the single unlocked exit and bodily searched, just in case they tried to take home a strand of thread or swatch of cloth. Communicable diseases were common for these work-ers crowded together in tiny rooms with little ventilation and no windows. Tuberculosis was known as "the tailors' disease," or "the Jewish disease."

On June 3, 1900, in response to the long hours, low pay, and dangerous working conditions in the garment industry, eleven delegates representing local unions in New York, Philadelphia, Baltimore, and Newark formed the International Ladies' Garment Workers' Union (ILGWU). The local unions banding together were the United Brotherhood of Cloak Makers, the Skirt Makers Union No. 1 of Greater New York, the Cloak Makers' Protective Union of Philadelphia, the Cloak Makers Union of Baltimore, the Cloak Makers' Union of Brownsville (in Brooklyn), and the Cloak Makers' Union of Newark, New Jersey. They were composed primarily of Jewish immigrants who had recently arrived from Eastern Europe, many of whom were socialists and had been active trade unionists before coming to America. The ILGWU was granted a charter from the American Federation of Labor (AFL) on June 22, 1900.

The decade that followed in the garment industry, especially in Manhattan, was turbulent and marked by wildcat strikes and other actions, as workers struggled for better wages and working conditions. In the summer of 1909, hundreds of tailors, buttonhole makers, neckwear workers, and waist makers from shops all across Manhattan walked off work on wildcat strikes. The strikes were short-lived, lasting briefly until the owners gave the workers a modest wage increase.

But in September 1909, events occurred that fundamentally changed the relationship between workers and their employers. Workers, mostly women, at the Triangle Shirtwaist Factory in Manhattan went out on strike. The manufacture of shirtwaists, or blouses, was a rapidly growing industry. New York had more than five hundred "waist" factories operating, employing more than forty thousand workers. The strike began when workers at Triangle overwhelmingly voted to join the United Hebrew Trades, an association of Jewish labor unions, rather than continue participation in the company-run "benevolent" association. Triangle's owners, Max Blanck and Isaac Harris, once "greenhorns" themselves, were now wealthy Manhattan industrialists, living with servants near the Hudson River at their in-town mansions. Triangle was a million-dollar business, the largest shirtwaist factory in Manhattan, with more than five hundred employees. Blanck and Harris responded to the strike by firing the union organizers and replacing them with prostitutes. In sympathy, other workers at Triangle joined the strike. Doubling down, Blanck and Harris hired gang members, who worked for Tammany Hall bosses, to threaten and assault the striking women, sometimes with the assistance of the New York police.

After five weeks of being out on strike, shirtwaist workers held a meeting and rally in the Great Hall at Cooper Union. Samuel Gompers, president of the AFL, was there endorsing the strike. But it was Clara Lemlich who carried the day. Barely five feet tall, Lemlich backed down to no one. At age twenty-three, she was a seasoned union organizer. The Triangle strike was her third strike in three years. In today's parlance, she was a "salt," an organizer who, unknown to the employer, gets herself hired solely to work inside the workplace for the purpose of organizing the workers. She had already survived broken ribs and other injuries in a brutal beating by a hired thug who had twice done time for burglary. On this night, following Gompers, she stood at the podium, her curly hair tightly pulled back and parted on the right, and forcefully declared in Yiddish, "I have no further patience for talk. I move that we go on a general strike! Enough is enough!" *Genuv iz genuv!* A general strike was called across the Manhattan garment industry, and workers from other sweatshops, again mostly women, walked off their jobs. Known as the "Uprising of the Twenty Thousand," it is believed that more than forty thousand garment workers participated in the strike called by Lemlich.

Once again, the owners turned the gangs loose on the strikers. But it was public opinion that turned the tide in favor of the workers. The images and stories of young women being harassed, threatened, and in some cases beaten, by gangs, company guards, and police, helped build public support for the strikers. And much of the support came from the unlikeliest of sources. Wealthy Manhattan women, many of whom undoubtedly owned the dresses painstakingly manufactured by the strikers, became their biggest supporters. These progressive socialites, who themselves were struggling for the rights of women's suffrage, found common cause with the immigrant women strikers. J. P. Morgan's daughter Anne announced in the *New York Times* her support for the strikers by joining the Women's Trade Union League:

> If we come to fully recognize these conditions we can't live our own lives without doing something to help them, bringing them at least the support of public opinion. We can see from the general trade conditions how difficult it must be for these girls to get along. Of course, the consumer must be protected, but when you hear of a woman who presses forty dozen skirts for $8 a week something must be very wrong. And fifty-two hours a week seems little enough to ask. . . . These conditions are terrible, and the girls must be helped to organize . . . and if public opinion is on their side they will be able to do it.[7]

Another supporter of the strike was Alva Vanderbilt Belmont, whose first husband was the grandson of Cornelius Vanderbilt. She divorced Vanderbilt as an adulterer and married August Belmont, whose father was a Jewish investment banker for the Rothschild family.[8] His mother was the daughter of Commodore Matthew Perry. By her birth and social status, she seemed the most unlikely benefactor of immigrant women garment workers. But after August Belmont died suddenly, Alva used her considerable wealth in support of women's and workers' rights, including financial support for the Women's Trade Union League. She also kept the socialist newspaper *The Masses* from bankruptcy. And in the "Uprising of Twenty Thousand," she gave considerable financial support to the strikers, including paying the bail of arrested strikers.[9] Anne Morgan, Alva Belmont, and others like them were bound together with Clara Lemlich, the small, skinny Jewish girl from the Ukraine. The right to vote, and the right to better working conditions and pay, were now united.

Their support proved pivotal, as the shirtwaist companies bowed to public pressure and agreed in early 1910 to negotiations and arbitration with the ILGWU. An agreement was reached on February 13, 1910, limiting the workweek to *only* fifty-two hours, while giving workers four paid holidays. In addition, employers were required to provide all the tools and materials necessary for the work, instead of charging workers for needles and thread. However, not all shirtwaist companies signed the agreement. Most notably, Blanck and Harris at Triangle did not.[10] Their refusal to move with the progressive tide would prove fatal to 146 of their workers a little more than twelve months later.

But, before that, there was more unrest. On July 7, 1910, more than sixty thousand cloak makers, this time mostly men, went on strike, affecting around eighteen hundred shops and stores across the nation.[11] With public opinion still on the workers' side, pressure to settle came from a new source—retail store owners led by Abraham Lincoln Filene, owner of the Boston-based Filene's. Filene, born of German-Jewish immigrant parents ten days before his namesake was assassinated, was unusual among the wealthy businessman of his time. He was an early supporter of women's suffrage, and much later bucked the trend of his business class and supported Roosevelt's New Deal. He was what some might call an enlightened capitalist. Although his strong personal beliefs guided many of the decisions in his life, he also had an economic self-interest in resolving the cloak makers' strike: he sold the coats and garments that they made, and without new merchandise arriving in his stores, his sales would fall off. So he would lead the effort to settle the strike, and sought the help of another Jewish Bostonian. He hired lawyer Louis Brandeis to negotiate and mediate a settlement.

Brandeis, born in Louisville, Kentucky, started his first law practice in Boston, after attending Harvard Law School, where he graduated first in his class. Graduating number two at Harvard was his partner, Samuel Warren. Brandeis soon became known as the "people's lawyer" for the many public-interest cases he handled, often for no fee. He was a brilliant lawyer and tactician, and later a United States Supreme Court justice. Filene also reached out to Meyer Bloomfield, a prominent Boston social worker and industrial reformer. Teamed with Brandeis, they were the perfect delegates for Filene to navigate the ever-widening gap between the cloak makers' manufacturers, who formed a protective association, and the striking workers and their union, the ILGWU.

Beginning in late July 1910, Brandeis and Bloomfield mediated and negotiated between the manufacturers and their corporate lawyer, Julius Cohen, and the union and its leaders. Brandeis attempted to craft a compromise, which early on the union had rejected and walked away from. This proved to be a tactical mistake, as for the first time public opinion began to turn against the union and the strikers, with the *New York Times* leading the way, writing that the workers were greedy and selfish. This setback made it harder for Brandeis to find a settlement. In the end, and after Cohen moved the strike into the courtroom, where he obtained an injunction against the ILGWU, a deal was reached on September 2. Dubbed the "Protocols of Peace," the agreement ushered in a new system of industrial relations. It was a watershed event in labor relations. It contained the common terms found in a labor contract today, covering wages, hours, and other working conditions, and was far better than most contracts of the day. Most important, and lasting into today, the protocols codified Brandeis's vision of industrial peace that would ban all strikes and lockouts during the term of an agreement, and replace strife with a grievance and arbitration procedure. Disputes would be resolved in arbitration and not on the streets. Finally, the protocols created the "preferential shop," which was, for all practical purposes, a closed union shop.

As 1910 came to a close, and with nearly 90 percent of all cloak makers now in the union and covered by the protocols, the manufacturers believed that the days of wildcat strikes and labor unrest were over. They could go back to making blouses and coats and other garments for the new American consumer, who was buying manufactured clothing at the rate of $1.3 billion in sales per year. In today's dollars, they sold $23 billion through stores like Filene's and in Sears, Roebuck catalogs.[12] Abraham Lincoln Filene, with Brandeis and Bloomfield, believed that they had created a template for industrial relations, built on mutual respect benefiting both capital and labor. Their peace, however, lasted only a short time. The revolution in industrial relations that the workers' strikes of 1909–1910 begat would continue following the Triangle fire, but at the cost of 146 lives.

2

THE TORCH THAT LIGHTED UP THE INDUSTRIAL SCENE

The Triangle Shirtwaist Company was located on the corner of Greene Street and Washington Place, just to the east of Washington Square, in the heart of Greenwich Village. Blanck and Harris, the "Shirtwaist Kings," moved their growing business from the dank tenements of Wooster Street to a modern, ten-story framed skyscraper with more than ninety thousand square feet of loft space to lease. The Asch Building was typical of the skyscrapers rising up all over Manhattan. In the first decade of the twentieth century, more than eight hundred new skyscrapers were added to the Manhattan skyline, many housing new loft factories.[1] The business model was simple. On the same-size footprint that the squalid tenement sweatshops occupied, modern factories boasting thousands of square feet rose one on top of the other. The home of commerce was going vertical.

Triangle leased the top three floors of the Asch Building. The added space was not for the benefit of the existing workforce. Instead, more workers were crowded into the new space. More of them were working in long rows of electric sewing machines and presses, all connected to a single power source, where they churned out more work, more waists, and all at faster speeds. For Blanck and Harris, and others like them, the high-rise factories enabled them to have all their workers under one roof. All their operations—production, shipping, and sales—were also now under one roof. The efficiencies of these modern factories allowed them to expand their product lines and make more garments than ever before. And, at Triangle, the "greenhorns" from Russia would make even more money than they ever imagined. In March 1911, all housed in the Asch Building, the Triangle empire was at its zenith. It was also ten floors above the streets of Manhattan, far above where the New York firefighting equipment could reach.

March 25, 1911, was by all accounts a beautiful spring day in New York. By late that Saturday afternoon, New Yorkers were out and about enjoying the weather around leafy Washington Square. At around 4:30 p.m., workers at Triangle were getting ready to go home. It has never been definitively determined how many workers were at Triangle that day. There were always some who were sick, while others chose to observe the Sabbath at home, but at the cost of a day's pay. Still, with sales people, office clerks, and management, including Harris and Blanck, and Blanck's young daughter Mildred, there were nearly five hundred souls crowded into floors eight through ten that day.

Production was on the eighth and ninth floors, while pressing, packing, shipping, and the executive offices were housed at the top, on floor ten. On floor eight, around 150 workers cut and sewed shirtwaists, all stationed at rows of long wooden tables. The cutting tables at Triangle were designed with boards surrounding the table legs, effectively creating wooden bins into which the cutters would toss cotton scraps and the tissue paper from paper patterns that hung on wires above the tables. Cutters and sewers would assemble anywhere from ten to twelve thousand shirtwaists each week, which in turn created a highly flammable mix of thousands of pounds of tissue paper and fabric scrap jammed into the wooden bins. The cotton fabric, which was more flammable than even the tissue paper, created a "virtual firebomb" on the eighth floor. The blaze was likely ignited by a still lit cigarette or match, carelessly tossed into a wooden bin.

Floor nine housed 278 sewing machines. Again, it is unknown exactly how many workers, mostly women, were at work that day. But, together with examiners, foremen, and bookkeepers, the best guess is that there were 250 employees on the ninth floor.

It was estimated that on the top floor of the Asch Building, on March 25, there were around seventy persons. Getting off the elevator on the tenth floor, one had to pass the reception area and switchboard. Beyond were the large offices of Blanck and Harris, looking down onto Washington Place through arched windows. Beyond the offices was a showroom where department store buyers and Triangle salesmen gathered to strike their deals. Finally, there were the pressing, packing, and shipping departments. The irons used to press the finished garments were connected by a web of tubing that supplied pressurized gas to heat them.

The exits from floors eight and nine were small, allowing only one person at a time to pass through a wooden partition. There, a guard would inspect the workers' handbags to make sure they were not stealing pieces of cloth. Still worse, the doors opened into, and not out from, the loft space. This was a fatal design flaw, as later, when jammed with bodies trying to escape, the exits were all but impassable.

From 4:45 p.m., when the first employee on floor eight yelled fire, and alarm box 289 was pulled, it took less than fifteen minutes for nearly 146 lives to perish in a firestorm. A few died later of their injuries. In that time, firefighters from New York engine companies converged onto the corner of Greene and Washington, including the tallest fire ladder from Hook and Ladder No. 20, but which still fell short of reaching the workers by some thirty feet. Lugging hundreds of feet of fire hose with them into the inferno, firemen bravely climbed the staircase as frenzied workers from the eighth floor fled past them to safety. Workers on floor nine were not so lucky. The exit doors were locked, or jammed, with the press of falling bodies. As many as ninety workers were hopelessly trapped on that floor.

At 4:50 p.m., the first worker jumped from the blaze on floor nine, falling to death on the pavement below. Soon, safety nets were deployed to catch the falling bodies, human missiles speeding more than one hundred feet down into the nets. Some of the bodies were on fire as they fell. A reporter on the scene wrote about "a new sound—a more horrible sound than description can picture. It was the thud of a speeding, living body on a stone side walk." Nonetheless, and probably emboldened by the nets, more workers jumped to their deaths, some two at a time, holding on to each other. Others struggled to remain vertical on the way down, believing that landing feet first might somehow save them. Three minutes later, the firemen stopped using the nets, and still the bodies fell, until the last worker jumped at 4:57 p.m. She was a woman who fell to her death only after becoming impaled on a hook holding up a sign on the side of the building. It is believed that in this brief time span, fifty-four workers jumped, choosing death from the fall rather than from the fire. This toll was in addition to around twenty-four workers who fell to their deaths while trying to flee by the fire escape, which was too small and too narrow for hundreds of workers to traverse in a fire. Still others tried to escape down the elevator shafts, burning the skin off their hands as they slid down the elevator cable before losing their grip and falling to their deaths. Eventually, nineteen bodies were recovered from atop the elevator car, the weight of which prevented the elevator from climbing to rescue more trapped workers.

The fire was finally contained on the top three floors by 5:15 p.m. In the end, 146 employees of the Triangle Shirtwaist Factory, mostly all women, perished in less than thirty minutes. Compounding the tragedy, there was never a reliable list of the victims, with many bodies burned beyond recognition. Still, many more were lucky enough to survive, including Blanck and Harris and others from the tenth floor who found refuge on the rooftop.

The fire did not have to happen. The Triangle fire occurred at a time when firewalls, fire doors and stairs, and automatic sprinklers had been commonplace in cotton mills since the 1880s. But although many cotton mill owners implemented fire safety measures because the cost of fires was too expensive, the garment industry had a different take on the financial efficacy of factory fires.

Four days after the fire, more than a half million New Yorkers lined the streets of Manhattan and marched in a funeral parade for the Triangle victims. Politicians and the press tried to assess blame for the fire. It was the fault of the building owner, the New York building department for allowing too-tall buildings with little thought for fire safety, and even the vaunted fire department. On April 12, just three short weeks after the fire, the politically ambitious district attorney, Charles Whitman, indicted Blanck and Harris for manslaughter. Whitman and the grand jury charged that the doors on the ninth floor were locked when the fire erupted. Blanck and Harris were arrested while at work at their new factory. For their defense, they hired Max Steuer, a former garment worker, who was known at the time as the greatest trial lawyer in America. On December 27, after nearly three weeks of trial and dozens of survivor witnesses, the jury acquitted Blanck and Harris on all counts after only a couple of hours of deliberations. The jury of twelve was made up of twelve businessmen, including a shirtwaist manufacturer. Blanck and Harris escaped the angry crowds camped outside the courtroom through the prisoner's exit.

After the fire, the upward trajectory of the Shirtwaist Kings collapsed. While they collected thousands of dollars from insurance, above the value of the inventory and machine losses, they were also forced to defend many civil lawsuits from the families of the victims. They turned to Steuer, who again successfully defended them from the civil claims. Blanck and Harris were never found guilty in the civil trials, and never paid a penny to the families. Yet they struggled to keep Triangle afloat. By 1918, the Triangle Shirtwaist Company was no more.

In terms of loss of human life, there were worse industrial accidents at the time of the Triangle fire. In 1908, 354 coal miners perished in the mines of West Virginia. But garment workers in Manhattan factories were not miners. They were cutting and sewing clothing in a modern office building. Although the work was grueling and the conditions were deplorable in many of the shops, working in a garment factory in Manhattan was not a known risk. The public response to this game changing event was swift and, once again, led by progressive forces from among New York's social elites. They lobbied at the capital in Albany, finding political support from Robert Wagner, a future United States senator, and Al Smith, a future governor and a presidential candidate in 1928. Together, they proposed legislation that created the Factory Investigating Commission. It was signed into law on June 30, 1911, just three months after the Triangle fire.[2]

The commission was empowered "to investigate the conditions under which manufacturing is carried on." Originally authorized for only one year, the commission lasted for four years. In addition to Wagner and Smith, the AFL president Samuel Gompers was a powerful member. Frances Perkins, who was a young Columbia graduate student from Boston when she heard the alarms for the Triangle fire while drinking tea nearby, was a frequent witness before the commission and a noted expert on workers' health and safety.[3] She was later appointed by Franklin Roosevelt as his secretary of labor, the first woman to hold a cabinet position.

In its first two years, the commission held fifty-nine public hearings throughout the state, with 472 witnesses providing sworn testimony on the working conditions of men and women in New York. Their testimony filled more than seven thousand pages of public records. The commission staff investigated conditions at 3,385 workplaces, including 359 chemical plants, which in 1912 made up 28 percent of all production in the United States. In the end, the commission's report in 1912 concluded that "health is the principal asset of the working man and working woman" and that it is up to the government to do everything in its power to protect the health and safety of American workers. Toward that end, the commission drafted twenty-six bills to protect workers' health and safety. From 1912 to 1914, thirteen of seventeen bills it proposed were enacted into law by the New York legislature. The New York commission was the most important investigative body in the United States up to then.[4] But it was not the only one. As Frances Perkins later noted, the Triangle fire was the "torch that lighted up the industrial scene."[5]

In May 1911, Wisconsin became the first state to enact workers' compensation laws.[6] Before the year was out, nine more states passed workers' compensation laws. At the end of 1913, eleven more states created workers' compensation systems. By 1921, only a decade after the Triangle fire, forty-six jurisdictions had workers' compensation laws. The incentive for such laws came in large part from employers. Up until then, workers could sue for damages in civil courts, but success was far from guaranteed. Initially, it was difficult to persuade a jury to lay blame for workers' injuries on the employers, who time and again successfully asserted contributory negligence and assumption-of-risk defenses, which shifted liability and responsibility to the worker. But as public opinion began to move against employers and in favor of injured workers, large jury awards became more frequent, and workers' compensation was born. A pooled insurance system financed by employer premiums and run by the state, and which substituted a fixed compensation system for the vagaries of jury awards, was in the end in the employers' best economic interests. Organized labor at first opposed workers' compensation as depriving workers of access to the civil courts to sue for unlimited damages. However, led by Gompers, labor eventually threw its support behind workers' compensation in 1909.

In addition to workers' compensation, other Progressive Era reforms included the idea of state industrial commissions enforcing worker health and safety laws. Conceived and promoted by John R. Commons, a professor at the University of Wisconsin, such commissions could establish regulations and guidelines by administrative rule making, without having to seek a legislative imprimatur every time health and safety rules needed to be implemented or changed. It took the politics out of workers' health and safety and put the latter into the hands of commissions with specialized expertise to develop rules and guidelines. The commission guidelines would make factory inspections clearer and more reliable for both employers and workers. Commons and his students drafted the Wisconsin Safe Place Statute of 1911, known as the "Wisconsin idea." When enacted, it created the first modern industrial commission with rule-making power. By 1936, administrative rule making would become the principal way states regulated worker health and safety. Yet the "Wisconsin idea" did not take hold as national legislation until 1970 and the act establishing the Occupational Safety and Health Administration (OSHA).

Finally, the Progressive Era produced the first investigations and reports on occupational diseases. The connection between work and workers' diseases had been widely known for decades. This was particularly true in New England textile mills, where poor ventilation and the inhalation of cotton dust led to chronic worker respiratory illnesses, including tuberculosis, while workers were also afflicted with skin diseases from the humid heat, deafness from the noise of weaving machines, and even cancers from carcinogenic lubricants. But it was not until 1911, when the State of Illinois published the results of a special occupational disease commission, that public awareness grew. Led by researchers and scientists from the University of Chicago, this commission identified thirty toxic substances commonly found in the workplace, including mercury, arsenic, and lead. The report's section on lead was the most damning, finding lead poisoning among workers in a wide range of industries, including smelting and refining, painting, printing, and battery making. In one enamel plant, the commission found 92 out of 148 workers "leaded." A version of the commission's recommendations became law in 1911, as the Illinois Occupational Disease Act, although it was watered down by business lobbying.

These Progressive Era reforms, in response to intensive industrialization, labor strife, and horrific worker deaths and injuries, were not without push-back from business interests and their politicians. In New York, just as the Factory Investigating Commission was experiencing its greatest legislative achievements, Governor William Sulzer, a Democrat and a big supporter of the commission, was impeached and removed from office. A weakened Democratic Party followed. Big business seized this political opportunity to charge that the commission was unfair and that it was sensationalizing the state of working conditions in New York, causing businesses to flee to more friendly jurisdictions and away from the heavy hand of government regulations. New York business lobbyists pushed the familiar refrains of big-business groups today, such as the U.S. Chamber of Commerce, and were successful in 1915 in vetoing all but one of the bills supported by the commission and its progressive allies. But, politics being what they are, fortunes change. In 1919, New York established an industrial commission with rule-making authority, naming Frances Perkins as a member.

The Progressive Era began an awakening in America about the collective responsibility for the health and safety of workers. It produced some of the most profound laws and ideas designed to carry out this social compact that America has ever seen. The historical figures of this time are unmatched since in their urgency and devotion to the idea that government is the only institution that can guarantee the health and safety of American workers, and that it has a duty to do so. The ideas born in the Progressive Era found new support from Franklin Roosevelt and the New Deal and from his secretary of labor. Social Security was created, giving workers needed financial security at the end of their working lives. And Robert Wagner, the senator from New York, sponsored the Wagner Act of 1935, which for the first time gave workers the rights to organize into unions, collectively bargain, and to strike. But these achievements would meet fierce headwinds then, and today, from powerful business groups and their lobbyists. The struggle for workers' rights, including the right to a safe workplace, was born in the red-hot forge of the Industrial Revolution and the Triangle fire tragedy. The successes of the Progressive Era and the New Deal still stand as signal legislative achievements for workers. American workers would still have to wait until 1970 before the federal government finally decided to protect all workers—more than sixty years after the ashes from the Triangle fire turned cold.

3

KEEPING AMERICANS SAFE AT WORK

Long before 1970, worker safety laws were a byzantine maze of state and limited federal laws that arose out of an alarming number of workplace injuries and fatalities as American businesses developed their unique system of industrial technology. After the Civil War, labor leaders promoted the creation of state labor bureaus, which collected statistics (such as they were at the time) and conducted investigations, but were without any legal authority to implement their recommendations. The nascent workers' safety movement that took hold in the second half of the nineteenth century was a confluence of "three styles of public policy": the regulatory approach of state bureaus and regulatory commissions; "voluntarism" practiced by Progressive Era reformers, which investigated and publicized dangerous working conditions; and the workers' compensation system, which raised the economic stakes for employers to create safe workplaces.[1]

On the regulatory front, in 1869 Massachusetts created the first labor bureau, the Massachusetts Bureau of Labor Statistics. By 1883, twelve more states followed the lead of Massachusetts. Even without any real authority, the reports of the state bureaus did much to influence public opinion and, eventually, state legislatures. Again, Massachusetts led the way. In its 1870 report, the Massachusetts bureau reported on the dangers from the lack of ventilation in factories, and called for legislation. Each year, it reported on deaths, injuries, and illnesses to workers in Massachusetts, providing fuel for reformers and early progressives who lobbied for worker protections. By 1877, Massachusetts passed the first factory inspection law, which required, among other things, fire exits and safety guards on machinery. Once again, other states followed suit. By 1890, nine states provided for factory inspections, and twenty-one states enacted some form of factory safety law.[2]

While these initial reforms were encouraging, they remained the status quo for many more years. Left to the states, worker safety laws were uneven at best, creating a race to the bottom, with businesses looking to locate in states with few or no worker safety laws. The fear of business flight to friendlier jurisdictions made it politically unrealistic for the passage of comprehensive worker safety laws at the state level. The political paralysis of elected officials to protect workers was also a legacy of the Industrial Revolution.

As such, legislative victories for labor and workers were sporadic and not very effective, despite the growing and more sophisticated body of scientific knowledge about the dangers of work. With muckraking journalists and fledging social scientists, there was no shortage of "voluntarism" publicizing and reporting on working conditions. Famously, there was the "Pittsburgh Survey" published in six volumes between 1909 and 1914.[3]

The Pittsburgh Survey studied and reported in great detail on the living and working conditions of workers in Allegheny County, Pennsylvania. The survey was sponsored by the Russell Sage Foundation, a philanthropic organization whose mission was to study social problems caused by the relationship between labor and capital, and to propose reforms and legislation. The Sage Foundation was funded by a $35 million endowment from the widow of a wealthy financial speculator named Russell Sage, who made his enormous fortune from investments and ownership in western railroads, and in the Western Union Telegraph Company. His widow used his wealth to empower and benefit workers.[4]

Pittsburgh was a perfect laboratory for research into the social problems caused by the enormous growth of capital created by the Industrial

Revolution. It was the prototypical new American industrial city. At the time, Pittsburgh was the fifth-largest city in the United States, the largest producer of steel and iron, and home to U.S. Steel. It was also home to tens of thousands of unskilled immigrants from the Baltic and Slavic states of Europe, who were pushed across the Atlantic by problems in their home countries and pulled to Pittsburgh by the promise of work in a modern industrialized American city. More than fifty researchers in the new discipline of social science spent two years studying and documenting the lives of these workers. Among the new social scientists was Elizabeth Beardsley Butler, a recent graduate of Barnard College. She focused her research on the living and working conditions of women. While doing her research and living among these workers, Butler tragically contracted tuberculosis and died.

In addition to Butler, a young social worker and photojournalist, Lewis Hine, worked with the researchers photographing Pittsburgh steelworkers and their families. Hine documented their lives and created some of the most indelible and iconic images ever recorded. Photographs of young girls standing among rows of looms; a muscular man tightening a steel bolt and seemingly becoming another cog in the machine; and an ironworker casually sitting at the end of a narrow steel beam, suspended hundreds of feet in the air. Hine's images, majestic in their own right, became a powerful visual record of the plight of workers in the new industrial America.

In the end, tables, graphs, and charts, including a "Death Calendar" marking the dates that workers died in Allegheny County on account of their work, as well as photographs and interviews, were combined into a groundbreaking work of social science research. The Pittsburgh Survey identified for the first time many problems that were plaguing an industrial America dominated by huge national corporations such as U.S. Steel. The sins of industrialization included worker exploitation, environmental despoliation, and civic corruption. Crystal Eastman's volume of the survey, *Work-Accidents and the Law*, became the basis for the establishment of workers' compensation laws, including in New York, where she drafted the law. Of work-related accidents, Eastman concluded:

> By industrial accidents, Allegheny County loses more than 500 workmen
> every year, of whom nearly half are American born, 70 per cent are workmen

of skill and training, and 60 per cent have not yet reached the prime of their working life. Youth, skill, strength,—in a word, human power,—is what we are losing.

Is this loss a waste? This is a question which Pittsburgh and every industrial district must answer. If it is merely an inevitable loss in the course of industry, then it is something to grieve over and forget. If it is largely, or half, or partly unnecessary,—a waste of youth and skill and strength,—then it is something to fight about and not forget.[5]

On the federal level, the prospect of national workers' safety legislation was even gloomier than in the states. Federal action on worker health and safety occurred in fits and starts. In 1884, Congress created a federal Bureau of Labor, as part of the Department of the Interior. Creating the Bureau of Labor was a small victory for organized labor in its push for a cabinet-level labor department, which would not come until 1913.[6]

In 1887, Congress created the Interstate Commerce Commission, partly in response to the thousands of railroad workers' injuries and deaths. In 1893, the ICC urged Congress to legislatively address the issue of railroad injuries. Congress passed what was called the "coupler bill," which prohibited dangerous practices involved in the coupling and uncoupling of railroad cars, often while moving.[7]

The mining industry had some of the earliest legislative successes, but only after thousands of miners perished; mine deaths in 1907 were at an all-time high of around 3,242.[8] From 1900 through the end of the twentieth century, more than one hundred thousand miners lost their lives while working in the mines. This figure does not count hundreds of thousands more who slowly died while suffocating from black lung disease. The first federal mine safety legislation, modest even by the standards of the day, was enacted in 1891 when Congress passed a law that established minimum ventilation rules and banned the employment in mines of children under age twelve. Around the same time, every important coal state had some type of mine regulations, typically focusing on ventilation rules in response to fatal explosions.[9] Later, in 1910, the Bureau of Mines was created, but its mission was scientific, and it had no regulatory power. Indeed, one of its inspectors, Herbert M. Wilson, told the Coal Mining Institute of America: "[The fact] that the Bureau will have no authority to enforce the adoption of its recommendations is not a matter of concern."[10] Not until

1941, after a period of horrific miner deaths, would federal inspectors be granted the authority to go down into the mines. Like most regulatory action, especially when it came to worker safety, there was a predictable correlation between preventable tragedy, public outrage, and legislative salve. The run-up to the 1941 federal legislation authorizing mine inspections was no exception. In 1940, ninety-one miners perished in Bartley, West Virginia, followed a couple of months later by seventy-three deaths at Neffs, Ohio. In 1952, federal mining legislation authorizing annual inspections, among other reforms, was preceded by 119 dead miners in an explosion at West Frankfort, Illinois.[11] The 1969 Coal Mine Health and Safety Act followed the deaths of seventy-eight miners in another explosion, this time in 1968 in Farmington, West Virginia. Still, one year later, on December 30, 1970, thirty-eight miners were killed in a dust explosion in the Hurricane Creek mine disaster, near Hyden, Kentucky; the lone survivor was blown out of the mine by the force of the explosion. More than one hundred children were left fatherless. The historic record is clear: it always takes dead miners to secure the passage of health and safety laws to prevent more dead miners.

The 1969 Coal Act was the most comprehensive piece of legislation of any of the previous federal attempts at reining in the coal industry and preventing more miner deaths. Although the Coal Act reduced the level of coal dust that miners could breathe, researchers have since reported an alarming resurgence of black lung. The prevalence of the most severe form of black lung tripled between the 1980s and the 2000s. In Kentucky, the disease prevalence was 9 percent from 2005 to 2009.[12] More legislation for mines and miners has passed, yet deaths from working in the mines go on unabated. The year 2010 was the worst for fatalities in twenty years, with deaths attributed to gas explosions and from moving equipment, root causes that were believed to be under control.

The law finally establishing the Department of Labor was signed on March 4, 1913, during the remaining hours of William H. Taft's administration and while Woodrow Wilson waited in the wings to take the oath of office. Before then, more than one hundred bills and resolutions had been submitted to Congress since the end of the Civil War dealing with a federal labor department. Taft reluctantly signed the bill as one of his last official acts, knowing that the Democrat Wilson and a Democratic Congress had the votes to override a veto. Still, he tried in his final days in office to bargain with Samuel Gompers. He would sign the bill if Gompers dropped legislation that prohibited the use of antitrust laws against labor unions. Gompers would not back down, and Taft signed the bill anyway.[13]

After the creation of the Department of Labor, the Bureau of Labor Statistics began to compile accident and injury statistics, first in the iron and steel industry, and later expanding its data to other industries. It was not until the 1930s that BLS was able to collect reliable and uniform injury data. Approximately twenty-three thousand workplace deaths were documented in 1913. The first Labor Department secretary, William B. Wilson, wrote that his guiding principle was that instead of feeding workers "into the maw of unhealthy occupations . . . the thing to do is to make the unhealthy occupations healthy." Achieving this goal was interrupted by World War I, as worker health and safety took a backseat to war production.[14]

In between the world wars, and under Frances Perkins, the Department of Labor created the Bureau of Labor Standards, the first federal agency charged with promoting worker health and safety. Laudable as it was, the Bureau of Labor Standards focused primarily on assisting the states with their health and safety laws. Comprehensive federal legislation was still decades away.

Apart from the changes in mine safety laws, for all other workers, the post–World War II economic boom that greatly expanded the American middle class diverted attention from the deaths and injuries to the workers who created this prosperity. In postwar America, workers died and were injured at alarming rates, overshadowing the need for badly overdue national legislation. Occupational deaths in the 1950s reached an annual high of 16,000 in 1951, before dropping back to 13,300 in 1958.[15]

While workers continued to die, other social issues occupied the attention of Americans. There was the civil rights movement, then the looming war in Vietnam. As often happens in politics, seemingly unrelated events finally helped propel worker safety to the national stage.

In 1968, President Lyndon Johnson was caught up in the quagmire of Vietnam and sought to change the narrative back to domestic matters, of which the Great Society programs still stand as his shining achievement. Among Johnson's legislative proposals was a comprehensive national health and safety law protecting millions of American workers.[16] Johnson cited "the shame of a modern industrial nation" where more than fourteen thousand workers were killed every year and another two million were injured at work.[17] It was a sweeping piece of federal legislation and created an obligation for employers to "furnish employment and place of employment which are safe and healthful." The proposed legislation included unannounced inspections, fines, blacklists from government projects, and even jail.

With the killing fields of Vietnam being broadcast into American homes nightly, the secretary of labor, William Willard Wirtz, used similar horrific imagery to rally support for the plight of workers during his congressional testimony. He challenged Congress "to act to stop a carnage" in the workplace that went unchecked because Americans "can't see the blood on the food that they eat, on the things that they buy, and on the services that they get." Wirtz and the Labor Department backed up his strong words with graphic pictures of injured workers in a pamphlet titled *On the Job Slaughter.* Wirtz's hardball tactics brought a strong response from businesses and the U.S. Chamber of Commerce, which objected to giving the federal government such broad powers, preferring instead to keep worker safety a matter best handled by the fifty states separately. Predictably, Johnson's bill failed, undone and weighed down by Vietnam, domestic unrest, and his failing popularity. It never even saw the light of day for a vote.[18]

After the elections in 1968, Richard Nixon took up the cause of worker safety, albeit in a version much watered-down from the Johnson bill. He proposed a board to oversee health and safety in the workplace, which was charged more with research and education and assisting the states and their laws than with investigation, enforcement, and prevention. Regulation was an afterthought. More significantly, the Department of Labor would be cut out from overall authority over the new laws. Organized labor opposed the Nixon version, and, not surprisingly, business groups and the U.S. Chamber worked for the passage of a neutered bill. The midterm elections in 1970 ushered in a more liberal Democratic Congress, which quickly took up the matter and passed, in effect, the Johnson bill. Nixon, mired like Johnson in Southeast Asia, and with reelection nearing, indicated he would sign the bill, ending the fight. Finally, on December 29, 1970, Nixon signed the Occupational Safety and Health Act of 1970 in a ceremony at the Labor Department. The Labor Department now had complete control through OSHA over the comprehensive federal legislation protecting and promoting the health and safety of all American workers.[19]

At last, there was a national law signed by the president. OSHA would use its specialized expertise to promulgate rules and regulations to finally protect American workers. Section 5 of the Occupational Safety and Health Act—the OSH Act—codified the social compact that, at last, would make Americans safe in their workplaces. Known as the General Duty Clause, it unequivocally states, "Each employer shall furnish to each of his employees employment and a place of employment which are free from recognized hazards that are causing or are likely to cause death or serious physical harm to his employees."[20]

The charge of the General Duty Clause is breathtaking in its scope and simplicity. It commands employers to provide workers a workplace that is "free" from hazards causing or "likely to cause death or serious physical harm." No ifs, and, or buts. Free from hazards means zero hazards.

No longer would workers have to submit to the vagaries of a crazy quilt of state laws. Uncle Sam would be there to protect and defend them. Or so it seemed back then, with the AFL-CIO president George Meany standing behind Nixon, beaming like a new father, proudly watching as the stroke of a pen gave birth to OSHA, fulfilling the hopes and promises of earlier generations of American workers who gave their lives for this day.

Also present at the birth of OSHA were business and Republican leaders, standing there shoulder to shoulder with Meany and the Democratic leadership, all arrayed in a smiling row behind a perspiring Nixon. To the uninitiated it looked like a genuine *kumbaya* moment. But the Republican leadership and their allies were smiling, no doubt, because they were already plotting and scheming to take down the OSH Act. It was almost one hundred years to the day from when Massachusetts passed the first factory inspection law to the passage of the OSH Act. Yet it took only a few short years thereafter for the business community to mount its campaign to strip the OSH Act of its promise, making it a stillbirth, while politicizing worker health and safety, all as the death toll and injury rate of American workers continued to mount.

No federal agency has been the target of more political tug-of-war than OSHA. From the beginning, Republicans introduced legislation aimed at defunding OSHA and limiting its enforcement and regulatory abilities, even to the OSH Act's outright repeal. The act came under immediate attack by the business community. In the 1976 presidential campaign, President Gerald Ford said that the business community would like to "throw OSHA into the ocean."[21] From 1973 through 1976, over one hundred oversight hearings dealing with OSHA were held before eight different Senate and House committees. In 1975, Congress began to pass appropriations riders to relieve employers of some of the OSH Act's requirements. In 1978, Congress nearly passed a Small Business Administration amendment that would have exempted 70 percent of employer worksites from coverage under the OSH Act. By 1979, OSHA was known to some, including Republican senator Richard S. Schweiker, as "the most despised federal agency in existence." Schweiker, along with other senators, including some Democrats, sponsored legislation to "reform" OSHA before it was repealed altogether.[22] In the end, the bill was defeated with the help of labor unions who argued that the "reforms" made workers more unsafe at their workplaces.

Under George W. Bush's Department of Labor, OSHA regulations and enforcement efforts were repealed, stalled, and ignored. During the younger Bush's two terms in office, the *New York Times* reported that OSHA issued the fewest safety standards in its history, while the only new significant health standard was ordered by a court judge. Efforts to list as health hazards silica dust linked to lung cancer, and construction site noise, all affecting more than three million workers, went nowhere. In 2003, ten years of regulatory work to create new health and safety standards for workers in danger of contracting tuberculosis, including hospital workers and prison guards, died a quiet death of regulatory inaction. The Bush administration instead preferred voluntary measures. Labor Secretary Elaine L. Chao, speaking to business owners in 2002, said, "There are more words in the Federal Register describing OSHA regulations than there are words in the Bible. They're a lot less inspired to read and a lot harder to understand. This is not fair."[23]

After the 9/11 attacks, OSHA under Bush unilaterally decided that it would no longer enforce health and safety regulations in the wake of terrorist attacks or other national emergencies. As a result, thousands of first responders who were responsible for the rescue, recovery, and cleanup efforts at Ground Zero in Manhattan were exposed to a highly toxic stew of carcinogens. More than fifty different types of cancers have been attributed to working at the World Trade Center cleanup site, as well as asthma, other respiratory illnesses, and post-traumatic stress disorders. Many of the illnesses, the workers and experts contended, could have been avoided, or mitigated, with proper training and protective respiratory equipment.[24] Indeed, rather than disable health and safety standards in national emergencies like 9/11, or Hurricane Katrina, which present some of the most dangerous and toxic work sites ever known, OSHA should have doubled its efforts to protect these workers from known workplace risks. But it did not.

The business community attacks were led then, and today, by its chief lobby, the U.S. Chamber of Commerce. On its website, under OSHA reform, the Chamber touts its legislative proposals that would, among other things, allow for employer "self-audits," so long as the legislation protects employers from the results of their own audits by deeming the audit report as a privileged work product and not discoverable by attorneys representing workers.[25] In other words, the fox will guard the henhouse and not be held responsible for eating the chickens.

From 2008 through 2016, the Chamber has spent more than $60 million a year on lobbying, including more than $120 million in 2009, 2010, 2012 and 2014.[26] In 2012 alone, the Chamber spent more than $35 million on federal elections, most of it against Democratic candidates.[27] Since Thomas Donohue became its president in 1997, the Chamber has gone from a business trade association to an openly antiunion organization with an aggressive antiworker agenda, so much so that many of its moderate business members, including some of the largest Internet tech companies, as well as old-line companies, have begun to express their discomfort with its hard-line views. Nonetheless, encouraged in 2010 by the Republican takeover of the House, the Chamber ramped up its lobbying efforts to deregulate health and safety protections.

While the financial scale of the Chamber's antiworker agenda is overwhelming, none of it is really new. Seemingly forever, employers and their lobbyists have argued against the big hand of government inspection and enforcement. As far back as the Massachusetts factory inspection legislation and the Triangle fire, businesses and lobbyists have railed against worker health and safety legislation with threats of plant relocations and political retribution. Business lobbyists insisted instead that voluntary health and safety programs were the best way to ensure a safe workplace.

In 2010, the Obama Department of Labor proposed a rule requiring employers to establish injury and illness prevention programs. This compromise (considered appeasement by some), of more self-regulation and less government inspection, was met with outright derision and opposition from the Chamber. Known as "Injury and Illness Prevention Programs," or I2P2, the rule was dubbed by the Chamber as an "extra from Star Wars" and met with complaints that it would require employers to do the unthinkable: "find and fix" workplace hazards that are not currently regulated. In effect, the Chamber argued that if there was a known but currently unregulated workplace hazard, the government could not require an employer to remediate the hazard and protect its employees. An employer can voluntarily do the right thing, but it cannot be required to do so. The business lobbying effort successfully delayed the rule, and in the 2014 fall regulatory agenda, I2P2 rule making was put on the longterm action list, meaning that it would not occur during the Obama administration.

Trotting out old canards, the Chamber and its allies continue to espouse a Dickensian view of the workplace. They claim that regulations weigh down the economy, hurt small businesses, and are job killers. The truth, instead, is that workplace injuries cost $250 billion annually to the American economy, most of the costs in the form of productivity losses that go directly to a company's bottom line.[28] Moreover, unhappy workers, worried about their health and safety, do not make productive workers. Meaningful reductions in workplace injuries add to a company's bottom line, create more jobs, and add value to the economy.[29] But meaningful reductions cannot be accomplished alone by voluntary compliance. There must a robust and independent government role in protecting American workers.

With the election of President Trump, however, a robust and independent OSHA is at risk. The Chamber's antiregulatory agenda reached a fevered pitch with the results of the 2017 election. Without the slightest filter for hyperbole, Thomas Donohue hailed the Trump administration's commitment to deconstruct the administrative state, writing: "After a relentless, eight-year regulatory onslaught that loaded unprecedented burdens on businesses and the economy, regulatory relief is finally on the way. The president and leaders in Congress have quickly made good on campaign pledges to rein in the regulatory state."[30] This pledge, finally called in after years of the Chamber's seemingly unlimited political largesse to American politicians, will undo OSHA regulations, among many others. The pledge secured by the Chamber will undoubtedly come at a cost to American workers and their families, and will break the promise of the OSH Act to keep American workers safe and healthy at their jobs.

4

JUST THE FACTS

Without a well funded and resourced OSHA, American workers in the twenty-first century remain as vulnerable to preventable injury and death as were the 146 victims of the Triangle fire more than one hundred years ago. For 2015, BLS reported that 4,836 workers were killed on the job. This is the highest annual total since 2008, and the total incidence rate remained almost unchanged at 3.38 per 100,000 full-time equivalent (FTE) workers.[1] Stated more clearly, 93 workers die each week on the job, or around 13 deaths per day. At the same time, approximately 50,000 Americans die each year from occupational illnesses, or 150 workers dying each day. In 2014, nearly 3.8 million work-related injuries and illnesses were reported. According to BLS, this was an incidence rate of 3.4 injuries or illnesses per 100 FTE workers, a slight decrease from a rate of 3.5 in 2013.[2] However, the BLS survey is based on employer-reported data, an inherently unreliable source, because most injuries go unreported. Nor do the BLS numbers include the self-employed, or the millions of workers misclassified as independent contractors. Accordingly, the true rate is estimated at around 7.6 million to 11.4 million work injuries each year.[3]

The death and injury/illness numbers are not evenly spread across all industries or geographic regions. Again, nineteen states saw an increase in nonfatal injuries from 2013 to 2014, while data for nine states was unavailable, including that for North Dakota, which is in the midst of an unprecedented oil boom. In 2014, North Dakota had the second-highest job fatality rate of any state, ceding to Wyoming the top spot from 2013, when North Dakota had fifty-six deaths largely attributable to the oil industry. Other lowlights from the 2015 BLS fatality report include that twenty-one states had an increase in workplace fatalities in 2015. Deaths from construction and transportation increased, whereas deaths from falls to a lower level accounted for nearly 40 percent of fatal work injuries in the private construction industry in 2015. Deaths to men increased again in 2015, and account for 92 percent of workplace deaths. In 2015, women, compared with men, continued to experience death by homicide at work by a higher ratio. Latino deaths on the job increased 12 percent in 2015 to the most since 2007. Latino workers are more likely to die on the job than any other demographic group at 4.0 per 100,000 FTE workers in 2015. Broken down by industry, fatal injuries in private construction increased in 2015, the highest since 2008. But the highest death rate belongs to the combined category of agriculture, forestry, fishing, and hunting.[4]

And, in the era of austerity budgets and sequestration, workers are at greater risk than ever before. In 2015, there were only 1,840 OSHA and state inspectors, down from the previous year. There were more than 140 million workers. As a result, what remains unchanged is that workplace inspection is almost nonexistent among the nearly eight million work sites under OSHA's jurisdiction. According to OSHA, there is only one federal or state inspector for every 74,760 workers. Because of this understaffing, OSHA has only enough inspectors to inspect workplaces once every 145 years.[5] OSHA's office in Houston is a good case in point. The Houston office is tasked with covering twenty-six Texas counties, which are home to more refineries than any other OSHA office in the United States. In addition, in 2014 Houston led the nation with the most new construction.[6] Overall, Texas led the nation in 2014 with 524 workplace deaths. With all this under its watch, the Houston OSHA office has only twenty-four investigators.

At the state level, the story is worse. There are only twenty-eight OSHA approved State Plans, and they all continue to flounder. OSHA monitors state OSHA programs, known as the Federal Annual Monitoring Evaluation (FAME). For fiscal year 2014, FAME reported overall state staff shortages, incomplete inspections, and a failure to follow up on OSHA directives, including whistleblower programs.[7] By way of example, in fiscal year 2014, Indiana OSHA allocated only sixty-three positions, fewer than the seventy positions required, and only thirty-eight positions were filled. According to FAME, the effect of understaffing in Indiana was a decline of nearly 50 percent in its inspection goal. Other deficiencies noted in the FAME report include not investigating all complaints, not documenting all fatalities, not responding in a timely manner to imminent-danger complaints, and not adequately conducting postaccident interviews.[8]

And despite the constant drumbeat from the business community about the cost of penalties when violators are caught, the average penalty in 2015 for a serious violation of the law was around $2,148 for federal OSHA, and $1,317 for state violations.[9] For worker deaths, the median OSHA penalty in 2015 was $7,000, and lower under state laws at $3,500.[10] Since 1970, there have been only eighty-nine federal criminal cases prosecuted for OSHA violations, resulting in a total among all defendants of eighty-nine months in jail. During this same time, more than 395,000 workers died on the job. In 2015, by comparison, 213 federal environmental prosecutions were initiated, resulting in 129 years of jail time and $200 million in fines and restitution.[11]

Many agree that these statistics, grim as they are, reflect an overall downturn in workplace deaths and injuries. Yet there is a genuine skepticism about the reliability of these numbers, which are used by all sides to make their points about the efficacy of more or fewer health and safety regulations. Just how much faith should we place in the statistics from BLS and other government agencies?

OSHA requires both the Department of Labor and employers to collect and maintain accurate records of workplace injuries, illnesses, and deaths.[12] Annually, BLS conducts a Survey of Occupational Injuries and Illnesses (SOII). It is supposed to be the nation's workplace health and safety "report card." Since 1992, the Department of Labor has touted declining injury and death rates among American workers, evidenced by the SOII. In turn, pointing to these numbers as proof positive that the rates of workplace injuries, illnesses, and deaths are in retreat, employers have argued and lobbied for less regulation and more voluntary compliance.

However, according to a report by the House Committee on Education and Labor in June 2008, there remains a "hidden tragedy" in the underreporting of workplace injuries and illnesses.[13] The report found that as many as 69 percent of injuries and illnesses are never picked up by the SOII.[14] One recent study concluded that at least one-quarter of all amputations are not reported to OSHA.[15] In late 2009, the Government Accountability Office, Congress's investigative and auditing arm, similarly reported that employers and workers routinely underreport workplace injuries and illnesses, and that BLS does not verify the accuracy of the SOII data.[16] Academic studies have concluded that, among the reporting problems, the BLS survey simply does not count all the workplace injuries and illnesses because it excludes large groups of workers from its data—as many as 20 percent of all workers.[17] This includes government workers, self-employed workers, and farms with fewer than eleven employees. Significantly, the survey excludes police and firefighters—workers with high injury and death rates. In addition, occupational illnesses are difficult to identify, particularly where there may be a long period of time between exposure and the onset of illness and disease. Finally, some employers are simply unaware of their reporting requirements.[18] In March 2016, OSHA reported that 50 percent or more of severe injuries are not being reported. Reasons for underreporting included "choosing not to report because they [employers] perceive the cost of not reporting to be low."[19] But apart from these explanations, there is an endemic structural problem of underreporting that skews the numbers and prevents OSHA, policy makers, politicians, and the public from knowing the real number of injuries, illnesses, and deaths in American workplaces. This in turn hinders any meaningful debate and any attempts to address the issue of workplace safety.

The House Committee and GAO reports, as well as academic studies, media reports, and testimony from workers, found several causes for underreporting, including a willfulness among employers to underreport, and fear from workers. The common thread for the existence of both causes is OSHA's reliance on employer self-reporting, which would be minimized, or nonexistent, if there were more OSHA investigators. Rather than providing accurate data of workplace injuries and illnesses, self-reporting provides employers with a strong incentive to underreport. By underreporting, employers can lower their workers' compensation premiums, make it less likely that they will be a target of OSHA inspectors, and increase their odds of winning lucrative government contracts. But how does underreporting occur?

There is widespread evidence of worker intimidation and harassment designed to suppress the reporting of work-related injuries and illnesses, particularly in industries with a mostly immigrant workforce, such as poultry and meatpacking, where workers are more easily exploited. Other causes of underreporting include the following:[20]

- Many workers who are injured on the job are required to take drug tests if they dare report their injury, even when there is no evidence that drugs or alcohol were a contributing factor.
- Injured workers are brought back for "light duty," even after a serious injury, in order to avoid lost time that raises an employer's workers' compensation costs.
- Injured workers are steered to company medical staff who are trained to minimize the seriousness of an injury that would otherwise be reportable. The GAO found that more than a third of occupational health practitioners reported that they are pressured by employers to provide insufficient medical treatment in order to hide, or downplay, the significance of an injury.
- No-fault absentee policies, which result in termination after accumulating a number of occurrences, include absences due to work-related injuries and illnesses. This directly results in employee underreporting.
- Safety incentive programs that provide monetary incentives to a group of workers for remaining injury-free encourages underreporting as a result of "peer pressure" from other workers whose prize is threatened by a coworker's report of an injury.

In response to the report of the House Committee on Education and Labor, which was supported by academic studies and worker testimony, the U.S. Chamber of Commerce trotted out its high-paid public relations team, including "testimony" from Baruch Fellner, an attorney at Gibson, Dunn, & Crutcher, an internationally known law firm. Fellner, an admitted "frequent critic" of OSHA law and policy, asserted that claims of underreporting were overstated. He said that the BLS statistics showing declining injury rates "are and must be substantially reliable."[21] His testimony concluded that "underreporting is minimal" and that "employers are doing a good and conscientious job" at reporting.[22] The Chamber's testimony to the committee was in stark contrast to the voluminous record of scientific empirical evidence.

Effective January 1, 2015, OSHA revised its record-keeping rule that updated the list of industries exempt from the reporting requirement because of their relatively low work injury and illness rates. Importantly, the revised rule expands the list of severe work injuries that must be reported to OSHA within twenty-four hours to include work-related in-patient hospitalizations, amputations, and loss of an eye. Work-related fatalities must still be reported to OSHA within eight hours.[23] Naturally, the revised rule was greeted by the U.S. Chamber with skepticism, claiming, among other things, that the new reporting requirement "will create confusion, burden and exposure to citations for employers."[24]

Notwithstanding the back-and-forth over the reliability of the statistics, one thing is still certain: workers are dying and getting injured in preventable incidents every day they go to work. A brief and admittedly unscientific survey reveals the breadth of the problem across industries more than a decade into the twenty-first century.

- A thirty-six-year-old oil worker in North Dakota was killed on June 20, 2016, after being struck by the boom of a crane. A week earlier, another oil worker was killed in North Dakota from an explosion and fire, while two other workers suffered third-degree burns over 70 percent of their bodies.[25]
- In October 2105, a fifty-six-year-old construction worker in California was killed when a retaining wall collapsed on him. He worked for a temporary agency—Labor Ready—that provided workers for ARCO National Construction. Cal/OSHA investigated and issued a $375 penalty against Labor Ready for not having documentation that the worker was trained. No citation was issued against ARCO, which ran the construction site.[26]
- In September 2012, a twenty-two-year-old carnival worker was killed on Long Island when he was struck in the head by a spinning ride at the Feast of Mother Cabrini Festival.[27] A year earlier, a lifeguard at Tampa's Adventure Island died after being struck by lightning while standing in shallow water evacuating patrons, because the water park did not follow procedures as the thunderstorm approached.[28]

- Tyson Foods was fined more than $142,000 when a thirty-seven-year-old worker at its Dakota City, Nebraska, plant died after a piece of equipment he was working on collapsed, crushing him.[29]
- A fifteen-year-old was killed in August 2012 after getting stuck in a winch while working on a shrimp boat off the Mississippi Gulf Coast.[30]
- On August 6, 2012, more than a dozen workers were almost killed when a vapor cloud sparked a massive fire at Chevron's Richmond, Virginia, refinery while they were working on a forty-year-old leaky pipe that had not been replaced in an earlier round of maintenance.[31]
- On July 28, 2010, in Mount Carroll, Illinois, three young males, ages twenty, nineteen, and fourteen, became trapped in a corn bin on a grain storage facility. They were sent in there without safety harnesses to break down clumps of corn, so that the stored corn could easily move down through the sump holes and onto a conveyor belt. Two of the young men died from corn drowning. Imagine quicksand made of thousands of bushels of compressed corn kernels. When this occurs, there is an intense pressure of the corn kernels against the rib muscles and diaphragm, literally preventing a person from being able to draw breath. In scientific terms, it is called compressional asphyxia. Victims also drown from suffocation, caused by inhaling the kernels into their nose and mouth. The third young man, who tried to save his friends, was rescued after more than six hours of being entombed inside the corn bin.[32] OSHA reported a record fifty-one grain bin deaths in 2010. These workers help produce the thousands of corn-based products in your supermarkets, as well as the ethanol that helps fuel your car.
- In the summer of 2014 alone, record heat and drought cooked many parts of the United States. Heat illness afflicted 2,630 workers, and 18 workers died from heat-related causes.[33] Despite the heat, OSHA denied a petition to establish regulations to protect workers from death and serious injury from acute heat stress.[34] With record heat predicted to become the norm, worker advocacy groups believed that the time was right to lobby for some standard or regulation to protect workers. Simple, commonsense, and inexpensive procedures, including adequate water supplies and more frequent breaks, were rejected, leaving farmworkers, construction workers, and others toiling outdoors at considerable risk.

So it goes, in the world of the U.S. Chamber, where underreporting is overstated, up is down, and night is day. The Chamber would challenge whether water is wet if it could gain a delay in regulating a "recognized hazard" that the General Duty Clause commands be eliminated from any workplace. And so it goes for millions of workers every day who remain the most vulnerable segment of society when it comes to their health and safety. No amount of Chamber spin and doublespeak, or political treachery from our elected officials, is too shameful for those who believe that workers are nothing more than replaceable parts in the machinery of wealth creation in America, instead of our mothers and fathers, sons and daughters, friends and neighbors. Workers, and their advocacy groups, lack the Chamber's limitless millions of dollars to influence policy and politicians. But they own compelling and tragic stories of preventable deaths, injuries, and illness. Just maybe, a public awakening from their stories will shame our elected representatives into action. Just maybe, the voices and images of workers will provide the undeniable proof and counterpoint to those who have attempted to create an alternative reality for more than one hundred years. Just maybe. But it is time for the voices of the survivors and injured workers to be heard.

5

STORIES

Grocery Clerk

> You know, it's one thing to have a girl in a bathing suit down on
> the beach, where what with the glare nobody can look at each other
> much anyway, and another thing in the cool of the A&P, under the
> fluorescent lights, against all those stacked packages, with her feet
> paddling along naked over our checkerboard green-and-cream
> rubber-tile floor.
>
> JOHN UPDIKE, "A&P"

Back in the days of the Wild West, out where the buffalo roamed, stores
in frontier towns sold dry goods and supplies. Coffee, tea, hides, nails.
The demand for store-bought produce or fresh meat simply did not exist.
Americans grew, or killed, most everything they consumed. But as our
cities began to grow, and as more of us lived in urban areas, owned no
land, and lived in tenements and other multifamily housing, the need for
dry-goods stores with a more diverse product selection grew. In 1859,
the Great Atlantic & Pacific Tea Company was established, and it would

become the first great supermarket chain. By 1880, it had more than 150 stores, "elaborate temples of tea and coffee." Between 1922 and 1925, now called just A&P, it opened seven stores a day, and by the mid-1940s it dominated the retail grocery market with more than sixteen thousand locations, much as Walmart dominates retailing today.[1]

The early grocery store was small, less than a thousand square feet, and sold only dry grocery items, canned goods, and other nonperishables. Fresh produce, meats, and bread were still bought from nearby greengrocers, butchers, and bakeries. The stores of those days were generally not self-service but staffed by a couple of employees in starched white aprons, who picked a customer's dry groceries from shelves behind the counter. In 1916, Piggly Wiggly opened the first self-service store in Memphis. Customers helped themselves to groceries stocked on shelves. In the 1920s, small regional supermarkets like Kroger, American Stores, and National Tea began spreading out geographically. In California, Safeway was created in 1926 by Charles Merrill of Merrill Lynch, through a series of mergers. By the 1940s, the major chains began closing and consolidating their smaller stores and replacing them with larger and larger supermarkets. The number of total stores was halved in some instances, while total sales remained the same, or increased in the larger stores. These new supermarkets now began to lease space to butchers, bakers, and produce vendors, often so seamlessly that the customer believed they were all part of the same company.[2]

The next big change that affected the supermarket industry was the postwar growth of America's suburbs and the emergence of our automobile culture. Stay-at-home moms would drive to their shiny new supermarket, sometimes two or three times a day. These stores were large, at around ten thousand square feet. By the 1970s, the discounter emerged, putting value over amenities and slick-looking stores. The major players followed the trend, some more successfully than others. Discounting began A&P's demise, ending in bankruptcy, a 92 percent drop in its stock price, and delisting from the New York Stock Exchange. Next, in the 1980s, supermarkets seemingly went back to the frontier days and now sold general merchandise along with groceries, all under the same roof, albeit now grown to more than thirty thousand square feet in superstores.[3] Once again, Americans could now buy coffee, tea, and nails (no hides), as well as tires, clothing, tools, and lawn furniture. Not stopping there, in the 1990s the superstore morphed into the supercenter warehouse, led by Walmart, Costco, Target, and, in Michigan, Meijer. These stores average more than 170,000 square feet, with 60 percent of the sales floor devoted to nongrocery merchandise. The average supermarket sells nearly thirty-nine thousand items, or SKUs, while the larger stores may top out at sixty thousand or more.[4]

What has become obvious is that a grocery store today has more in common with a factory than it does with the iconic A&P of Updike's youth. Gone are the days when the meat department closed at 6 p.m., and the store was closed on Sunday. Shopping patterns changed as women entered the workforce and shopped after work. Sunday became the busiest day of the week. As a result, supermarkets today are large industrialized workplaces. The typical grocery store is open 24/7. More and more employees are needed to work in larger and larger stores, and to work at faster and faster speeds. A supercenter can employ as many as five hundred workers over three shifts. However, as companies try to squeeze out as much profit as they can, and with labor costs being the easiest to cut, many stores are cutting hours and the number of employees while still trying to maintain same-store sales. Doing more with less falls on stressed-out workers and creates a dangerous workplace. Even customers are being asked to pitch in by checking and bagging their own groceries. Stocking shelves and checking out groceries are now assigned production standards. A checker must scan and bag more groceries at faster speeds than ever before or face discipline for underperformance.

Musculoskeletal disorders (MSDs), particularly among cashiers, and repetitive trauma cases account for 73 percent of the total number of injuries in the industry. With the advent of scanners, cashier MSDs from the same repetitive motion of scanning merchandise across the electric eye more than six hundred times an hour has led to an epidemic of hand, back, neck, and shoulder injuries. Carpal tunnel, rotator cuff, back strains, and trigger finger are the names of just a few common grocery store injuries. Truckloads of merchandise arrive daily, all day long, that must be moved from the truck to the back room, into coolers and freezers, and finally onto the sales floors by hand jacks, forklifts, and other motorized assistance, driven by workers whizzing and dodging around co-employees, vendors, and customers. In the biggest stores, the height of shelves can now tower with skids of plastic-wrapped product precariously stacked near to the ceiling, above unwitting customers and workers. Big powerful balers are used to crush and bale cardboard and other packing materials. Meat and deli clerks stand on slippery floors while using knives, grinders, slicers, and saws.

All this occurs daily and before our eyes, creating a dangerous workplace for teens, women, and men. And the response from Congress and President George W. Bush was to reject ergonomic standards developed by OSHA and instead issue voluntary guidelines in 2004. Bowing to pressure from business and the U.S. Chamber of Commerce, OSHA took pains to emphasize that "these [voluntary] guidelines are advisory in nature and informational in content. They are not a new standard or regulation and do not create any new OSHA duties." In other words, ignore them without any consequences to the employer. The supermarket today is a long way from the bucolic country general store, or Updike's romantic A&P. It is a dangerous place.[5]

Hanna Phillips

If you only looked at Hanna Phillips you might confuse her with someone who could have been a high school cheerleader. Maybe even prom queen. She is slim and petite and has eyes that demand your attention when she looks at you. Just looking at her, you wouldn't know that behind her midwestern prettiness is a twenty-something who has literally been through life's grinder. Yet she blames no one or anything. She calls herself "lucky" without even the slightest hint of irony.

Hanna is upbeat and positive despite the Dickensian arc of her life. At eleven she was placed on probation by the juvenile authorities. Hanna acknowledges that she "got in trouble," but offered and recalled no further details other than to say that she was "incorrigible." When she was thirteen, the state found her mother unfit and placed Hanna in White's Residential & Family Services, a Christian residential home in Wabash, Indiana. At White's, Hanna worked at the canteen and was paid two dollars an hour. Hanna has had infrequent contact with her mother since then.

When she turned fifteen, Hanna was placed in a foster home in Wabash. There she met her best friend, another girl in foster care. While living in the foster home, Hanna was hired by the Kroger Company at its store in Wabash. Founded by Barney Kroger in Cincinnati in 1883, Kroger is one of the world's largest retailers. In 2011, it had nearly twenty-five hundred supermarkets operating under more than a dozen banners in thirty-one states, and had sales of over $90 billion.

Hanna was sixteen when she started at Kroger. She went to high school in the day and worked in the afternoons and evenings bagging groceries and pushing shopping carts. She had little time for other teenage activities. After a while, Hanna was promoted to cashier and dairy stocker. She worked hard and did jobs that others refused. One time she was told to change ceiling lights, at a height of nearly twenty feet above the hard concrete floor. With no safety harness, Hanna climbed to the top of the ladder and changed the fluorescent lights on the orders of her store manager.

OSHA reports that annually there are nearly twenty-five thousand accidents and thirty-six fatalities due to unsafe conditions and practices involving ladders and staircases. Proper training is required in all instances. Hanna had no training from Kroger. She was told to change the lightbulbs, and she responded as most workers do: she followed the directive of her store manager. She climbed the ladders holding eight-foot-long glass tubes, because insubordination was a fireable offense and because she wanted her manager to know that she was a hardworking employee.

For all her hard work and risk taking, Kroger kept Hanna's hours right below full time so it would not have to pay her any benefits. In her senior year of high school, Hanna turned eighteen and moved out of the foster home to live in an apartment on her own. She was able to work more hours at Kroger as part of a special program that allowed her to go to school in the morning and work the rest of the day. Kroger was hiring meat clerks and asked Hanna if she wanted to get more hours and work in the meat department. She said yes. Hanna described living on her own, supporting herself, going to school, and working full time as a pretty stressful time in her life. She had a lot to balance.

Up until the late 1980s, most retail supermarket meat departments were staffed by highly trained and skilled journeyman meat cutters. These meat cutters, mostly men, processed quarter sides of beef. The "hanging weight" of these sides of beef was around 150 pounds. They hung in the store coolers waiting to be processed into porterhouse, T-bone, sirloin, rib steaks, and other specialty cuts. In addition, these meat cutters processed whole pork into ham, spareribs, and bacon. Over time, much of this highly skilled processing has been outsourced to the slaughterhouse, where less-skilled and lower-paid workers process the beef and pork at factory speeds for delivery straight into the supermarket meat cases.

Consequently, most of the work today in a modern supermarket meat department is done by meat clerks, who are less skilled and lower paid than the journeyman butcher. Generally, and at Kroger, meat clerks stock the retail meat cases and clean the meat department at night. They still do some minimal processing, including grinding meat. It is not uncommon for a manager to instruct a meat clerk to actually cut meat for a special customer order, even though the clerk is not trained or paid to cut meat.

In a 2007 publication, OSHA reported 8,450 nonfatal workplace amputations for all private industry. In a supermarket meat department, meat slicers and grinders have contributed to this grisly toll. The Department of Labor has designated operating meatpacking and processing equipment as one of only six nonfarm jobs that is so dangerous that employees under eighteen are prohibited from working on these pieces of equipment. Still, when it comes to workplace amputations, OSHA has issued only *voluntary* standards to employers on how to prevent serious injuries to their employees caused by amputations. Safeguarding methods for meat grinders include on-the-job training and providing employees with properly sized plungers to eliminate the need for their hands to enter the throat of a grinder. Employees are also trained in an "energy control program." Known as LOTO, for lock out / tag out, these procedures are designed to shut down and actually lock out the power source of dangerous machinery, like grinders, during maintenance and cleaning.

A meat grinder is generally configured on the top with a stainless steel tray from which the employee feeds the meat down the throat of the grinder and into the auger area. From there it comes out as ground beef. When cleaning a grinder, it is important to clean out the throat and auger thoroughly. Bits and pieces of greasy meat get stuck on the wall of the throat. In that event, workers often have to reach their hand into the throat of the grinder to remove the small chunks of meat.

On April 8, 2009, Hanna was scheduled to work from 11:30 a.m. until 8 p.m. As the last person in the meat department, she was responsible for cleaning the processing room and equipment. Meat cutters and clerks alike wear long white lab coats over their clothing. At only a few inches over five feet, and around 110 pounds, Hanna was diminutive inside her lab coat. She had to roll up the sleeves two or three times just to free her hands. At around 6:30 that evening, Hanna started to clean the machines. She reached into the throat of the grinder to push the auger out for cleaning. She believed that the power source to the grinder was shut down. But the grinder was not LOTO. The power should have been disengaged, but it wasn't. She later learned that a maintenance person from the company that serviced the equipment manually overrode the switch that turned off the power. No one told Hanna. When she reached in with her right hand, her sleeve from the too-big jacket caught and pulled the power lever, turning on the grinder. After only a few seconds, Hanna pulled her arm out of the grinder. Her right hand and arm just below her elbow were gone, ground up in the meat grinder, leaving Hanna with only a bloody stump.

She was the only one in the meat department at the time. Hanna said she was "lucky" that she did not pass out. She would have bled to death because no one would have found her lying on the meat room floor. Hanna wrapped up her bloody stump in what was left of her meat coat and calmly walked to the front of the store for help. She was not crying or screaming but said that it felt like her arm was on fire. Hanna recalled that the store came to a dead stop. Customers and employees stopped what they were doing. The pharmacist tried to put bandages on what was left of Hanna's arm to stop the bleeding. Hanna sat down on a bench at the front of the store and waited for the ambulance to arrive. She said that she was not in shock. Rather, her adrenaline kicked in and she did what she had to do to get help. That was all she could do.

She was taken by ambulance to the Wabash County Hospital. Her injuries were so severe that she was airlifted by helicopter to the Parkview Hospital in nearby Fort Wayne, where doctors performed surgery for many hours. Hanna remained there hospitalized for another two days before being released. Back home she moved in temporarily with her then boyfriend's parents. She was on a lot of medication for pain and to protect her against infection. A couple of weeks after the surgery, the stitches came out, and after a while the swelling in her stump subsided.

Physical therapy followed, to teach Hanna how to use her prosthetic hand. While in physical therapy, she met a man who had been seriously injured a few years earlier from a fall at work. He is all in one piece despite the fall, but has been in constant pain since. Hanna said that compared to him, she was "lucky." Although she still has occasional pain, Hanna mostly has phantom feelings where her right hand and arm used to be.

She gets a tingling feeling where her hand once was, as if it were trying to wake up from a deep sleep. At other times she forgets that she lost her arm. Doctors are not sure why a person who lost a limb still reports the feeling of having the amputated limb. They speculate that the brain has not yet caught up with the reality of a lost limb.

The grip of the prosthesis is controlled by nerve impulses coming from her stump. Sometimes Hanna cannot control the nerve impulses that control her new right hand. When that occurs, the hand acts without her direction. One time when she rested her prosthetic hand on her leg it suddenly gripped her calf and would not let go. While this may evoke the scene from *Dr. Strangelove* where the title character's prosthetic arm involuntarily makes the Nazi salute, Hanna's experience was not at all humorous. It was very painful and frightening. As a result, Hanna doesn't trust her prosthesis and wears it less and less. She is learning to do more and more with just one arm.

Hanna could no longer work as a meat clerk, and for that she received $160 a week in workers' compensation benefits from the State of Indiana. Workers' compensation benefits in Indiana are among the lowest in the country. To make ends meet, she took on odd jobs with friends. After a while, she did not want to be a burden to her friends and, looking for a change of scenery, Hanna moved to Laguna Niguel in California. There she worked as a nanny taking care of a toddler boy. She enrolled in a nearby college and took some classes and began to earn college credit. She stayed in California for almost two years before returning home to Indiana. She enrolled at Ball State University in Muncie and continued taking classes, hoping to graduate with a degree in psychology. She left Ball State after a short time because "it was too much of a party school."

Hanna said that she never let herself get mad about the accident. She didn't want to waste time on something she couldn't fix. Even so, it took "a little bit to allow myself to cry." A lot of people advised Hanna to go on Social Security disability and to sit back and collect a check. Hanna said no, over and over again. "I can still work," she said. "I can still get a job." She did not consider herself disabled and said her biggest challenge was putting up her hair. There was not much else she could not do.

Hanna enrolled with a temporary worker service hoping to find full-time work. She revealed her lifting restriction. This made it hard for her to find work. Not many employers were willing to hire a one-armed person.

Under the Americans with Disabilities Act, employers cannot ask job applicants to answer medical questions, take a medical exam, or identify a disability. An employer may not ask a job applicant to answer medical questions or take a medical exam before extending a job offer. An employer also may not ask job applicants if they have a disability, or about the nature of an obvious disability, like having only one arm. Only after a job is offered to an applicant does the law allow an employer to condition the job offer on the applicant answering certain medical questions or successfully passing a medical and physical exam.

So Hanna would have been well within her rights not to reveal the nature of her disability to the temp agency. She only wanted the opportunity to demonstrate that she could perform the job. As a result, the temp agency didn't call very often. She blames herself for telling the temp agency the truth about being one-armed.

Her "big break" came in October 2012 when she got a call that Honda was hiring at its parts warehouse adjacent to a manufacturing plant that makes Honda Civics and Acuras. After offering her a job, Honda conditioned the offer on passing a physical exam. Hanna had to lift a series of increasingly heavy containers filled with salt pellets, then move the containers from one table to another. At first, she struggled lifting the container while wearing her prosthesis. It was too cumbersome and she did not trust its grip, so she took it off. With one good arm and her stump, Hanna successfully moved the containers and passed the physical test.

Her first job at the parts warehouse was to lift empty containers onto a conveyor belt and empty the trash. Hanna worked without her prosthesis. After a while, she was trained to drive a forklift to get parts from the warehouse and place them into containers that move on a conveyor belt over to the manufacturing plant. It is a fast-paced job, with lots of forklift drivers, mostly men, whipping around tight corners and down narrow aisles. It is like a bumper car ride at an amusement park, but far more dangerous. Accidents from crashes easily happen, and if the forks are raised, the accidents can be fatal. Hanna is not as fast as the other drivers because she has to steer, stop, and operate the forks all with one hand. The parts baskets are made of steel, and if they fall while she is picking them off the shelves, the baskets can crush her sitting in the forklift. Nonetheless, Hanna likes the job. She was "lucky" to find work with Honda, even if it is a nonunion plant paying her only $10.50 an hour and no health insurance.

After a while, Hanna contacted lawyers to see if she had a case for losing her arm in a grinder. Most lawyers she talked to told her that there was nothing they could do to help her. They said her damages against Kroger were limited to workers' compensation benefits through the state because she was injured at work. Hanna eventually found a lawyer who sued for negligence the company that serviced the grinder. After a few years of litigation, they settled on terms that could not be disclosed. Hanna said only that she was "lucky."

Hanna, at the time of this writing, is pregnant. She is going to have a boy. She is excited to be a mother. Her biggest challenge, she said, will be to change a squirming baby boy's diaper. She did it in California when she was a nanny and said she might have to use her feet to help hold him still. She will not use her prosthesis for fear that it might involuntarily grab hold of her baby. She told me of a YouTube video she has watched numerous times about a mother with no arms and legs caring for her baby. "If she can do it, I can do it one-handed," Hanna said. "I got hurt. But I got lucky."

Lori Keen

The back room of a typical grocery store is a crowded warehouse, chock-full of unstocked merchandise, eventually bound for all departments in the store. The back room is a relatively small and confined area with narrow aisles, because most space in a grocery store is devoted to selling the merchandise. Grocery stores do not make money in storage and warehousing. Semitrailers deliver the merchandise, in boxes, crates, and on shrink-wrapped wooden pallets, to the loading docks, where it is offloaded by employees driving forklifts and hand jacks. From there, workers stack the boxes and pallets of merchandise vertiginously one atop the other, where it remains until employees move it out onto the sales floor for stocking on the shelves. The backroom is also where the coolers and freezers are located. Garbage dumpsters and balers, huge machines where the hundreds of emptied boxes and other shipping material are compacted and baled for recycling, are also crowded into the back room. And, finally, the back room is where schedules are posted, along with government notices and store information, and where employees and management pass through to and from store offices, time clocks, bathrooms, employee break rooms, and lockers. The back room is a hive of activity, a messy, seemingly disorganized space, where every inch is crammed with merchandise, equipment, and supplies. It is an accident waiting to happen.

Lori Keen worked for the Kroger Company for thirteen years at its store in Franklin, Indiana. At the time of her death, Lori was working in the receiving dock area of the store, where beer, pop, and water vendors make their deliveries. On March 15, 2010, Lori was attempting to move two pallets of Ice Mountain bottled water with a heavy-duty forklift. Lori was double-stacking the pallets. Each pallet was stacked six rows high with cases of bottled water, and weighed, exclusive of the wooden pallet, approximately 2,050 pounds. Using a forklift known as the Crown Stacker, Lori raised the first pallet and lowered it on top of the other. This is known as "straddle stacking." The combined weight of the two pallets was over 4,000 pounds. The shrink-wrapped pallets now stood twelve rows high with bottled water, or more than fourteen feet tall when including the wooden pallets. With the pallets double-stacked, Lori used the Crown Stacker to lift the two pallets and move them toward the corner of the dock area for storage. The maximum load capacity for the Crown Stacker was 4,000 pounds. At some point during this process, the top pallet began to lean forward, and cases of the bottles on the bottom began to "crisscross," giving in to the weight of the load. At this moment, a Pepsi-Cola delivery driver entered the back room and warned Lori about the way she was stacking the pallets. He advised her to remove the top pallet with the forklift. He left the back room, and Lori continued to push the pallets across the room. As she did, the pallets began to go into a deeper lean. Lori stopped the forklift, got off, and walked over to the pallets on the side toward which they were leaning. The store surveillance video shows Lori with her arms raised in the air. Inexplicably, she placed her arms and slight weight against the 2,050 pounds of bottled water, which was leaning above her. What she was thinking is hard to say. Lori had to know that she could not push the cases back into alignment. But she was just trying to do her job the best that she knew how. While she was standing there pushing, both pallets collapsed, burying Lori Keen alive beneath more than 100 cases of bottled water, which weighed around 4,100 pounds. A Frito-Lay delivery driver working in the back room reported he heard "a noise like thunder," and then he heard Lori scream. He looked over and saw the top pallet coming down on top of her. The top pallet was mostly intact as it fell, while the bottom pallet was coming apart. The bottles in the bottom pallet had shifted and ripped through the shrink-wrap, sending the top pallet tumbling on top of Lori.

Post-accident photographs show a tangle of shrink-wrap, and mostly intact cases, heaped into a pile several feet high. On the bottom of the pile lay Lori Keen. She was pinned between the bottom pallet at her back, and the top pallet, which fell on her. The Frito-Lay worker reported that Lori was in a "crouching position, with her head between the pallets and one hand free."

The Franklin Police Department responded to the scene. After many minutes, coworkers were able to unbury Lori from the cases of water. She was unresponsive and was transported by ambulance to nearby Johnson Memorial Hospital. Emergency medical personnel determined that she had multiple traumas, including hypoxic brain injury, which occurs when oxygen stops flowing to the brain. She was then transported to Methodist Hospital in Indianapolis for further evaluation. Her condition did not improve, and doctors there concluded that she had suffered severe anoxic brain injury during the accident; her brain received no oxygen. As a result, she was pronounced "non-viable" on March 20, 2010, at 11:20 a.m. The Marion County coroner performed an external exam and reviewed the medical records. According to the Coroner's Verdict Report, Lori Keen died of "mechanical asphyxiation due to crushing beneath weight." The manner of her death was ruled an "accident."

Indiana OSHA (IOSHA) investigated the "accident" and cited Kroger for violating several OSHA regulations. Specifically, IOSHA found that Kroger did not properly train Lori and others operating forklifts on safe operations, including "load manipulation, stacking, and unstacking." During the opening investigative conference on March 15, the store manager told IOSHA that employees are not trained on straddle stacking. Then, when asked why the pallets of bottled water were stacked on top of each other, she answered, "It's what we've always done." IOSHA deemed this a "serious" violation and proposed a $3,500 penalty. Kroger was also cited because the bottled water was stacked, limited in height, "so that it was not stable and secure against sliding and collapse." This was also determined to be a "repeat" violation, because Kroger was cited for this violation in 2008 at another one of its stores in Georgia. The proposed penalty for the repeat violation that killed Lori Keen was $12,500. Finally,

Kroger was cited for a "nonserious" violation because loads were being handled that exceeded the capacity of the forklifts. The proposed penalty for this violation was $1,000. In total, Kroger was fined $17,000 for Lori Keen's preventable death. Kroger disagreed with IOSHA and contested all the citations and penalties. IOSHA held a conference with Kroger's risk manager and its attorney to discuss the proposed penalties. No one from Lori's family was invited to attend the conference. In the end, the safety orders were upheld, but the penalties were reduced to $8,000. In 2010, Kroger reported to the Securities and Exchange Commission net earnings of $1.1 billion. Beyond the penalties, Kroger was shielded from further liability by workers' compensation laws, despite its failure to properly train Lori.

Lori's husband, Billie Keen, filed suit against Nestlé Waters North America Inc., the manufacturer and distributer of Ice Mountain bottled water. The lawsuit alleged that Nestlé was at fault in the design of the plastic bottles. An expert witness testified that the accident was caused by the "down-sizing of the plastic bottle and product/package system used by Nestle to a point where it was no longer able to support the weight of a typical pallet during normal material handling." The result was "a bottle that is so thin and weak that it will not withstand normal stacking during storage and handling." Among the recommendations to improve the safety for those that handle the bottles of water, Keen's expert suggested an inexpensive warning sign placed on the pallets simply stating: "Do Not Stack." Other recommendations included safe handling instructions to the employees of Nestlé customers, like Kroger.

Nestlé's lawyers moved to dismiss the lawsuit and argued mainly that it did not have a duty to warn Lori Keen of the possible dangers of handling its pallets of bottled water. The federal court judge denied the motion, finding that it was reasonably foreseeable that if Nestlé did not use reasonable care in the design and packaging of its pallets, and failed to provide adequate warnings to end-users like Lori Keen, "the people handling the pallets may be subject to injury." The case eventually was settled before trial for an undisclosed amount.

Some criticized Billie Keen for filing the lawsuit against Nestlé, even going so far as calling it "frivolous." But, for Billie, he wanted someone to take responsibility for Lori's death. She went to work on March 15 fully expecting to return home at the end of her shift. She did not intend to be crushed to death by more than four thousand pounds of bottled water. She left behind a husband and a young daughter, her mother, and many friends and coworkers.

IOSHA deemed her death worth only $8,000. Workers' compensation shielded Kroger from liability. Lori was only thirty-two years old when she died a preventable and horrific death. She had a long life to live with her family and friends. Her obituary described some of the things that Lori enjoyed: decorating cakes, crocheting, and collecting dragons. But mostly she enjoyed spending time with her family and friends. She loved people.

Hotel Housekeeper

> I like to work half a day. I don't care if it's the first twelve hours
> or the second twelve hours. I just put in my half every day.
> It keeps me out of trouble.
>
> KEMMONS WILSON, FOUNDER OF HOLIDAY INN HOTELS

It was postwar America, and the love affair with our cars was made all the more intense when President Eisenhower championed and signed the Federal Aid Highway Act of 1956, which over the next five decades created more than forty-seven thousand miles of interstate highways. It was properly touted as the largest public works project since the pyramids. With more leisure time and money at their disposal, Americans headed for the highways, and they needed someplace reliable and clean to stay. In stepped Kemmons Wilson, who in 1952 had opened his first Holiday Inn of America in Memphis. It was marked by a large neon-green sign topped with a glowing star and the ubiquitous glowing red letters, "vacancy." Simmons called his motel chain Holiday Inn, after the movie of the same name starring Bing Crosby.[6]

But before Wilson, there was Conrad Hilton, who purchased his first hotel in 1919 in Cisco, Texas. The first roadside motor lodge, or auto court, opened in San Luis Obispo, California, in 1925, and charged tired travelers $2.50 a night.[7] The word *motel* is a combination of motor and hotel, because back then auto courts were located on the roadsides of America's early highway system, iconic roadways like Route 66. Travelers slowed down and eased their Packard or Nash into a gravelly parking space extending only a few yards between their motel room and the high-way traffic. In the recesses of my memory, traveling by car or van with my family in the '50s and '60s seemed both exotic and simpler than it is today. It was exotic to pull into a Holiday Inn, a Ramada Inn, or a HoJo, somewhere off the interstate on our way to or from Chicago. The rooms offered air conditioning, first introduced at the Hotel Statler in Detroit in 1934. There were color televisions in every room, and an in-ground swimming pool waiting for us just outside the door of our room and across the hot parking lot pavement. The rooms were clean, certified so by a paper strip across the toilet declaring that the seat was "sanitized for your protection." How comforting that must have been. Those nostalgic days are long gone.

According to the American Hotel Lodging Association, for year-end 2014, in the United States there were more than 53,000 properties and five million guest rooms. More than 6,000 of those properties offer more than 150 guest rooms, and of that number 537 hotels have more than 500 rooms. The lodging industry in 2014 generated $176 billion in sales. Most significant, average occupancy rates in 2014 were 64.4 percent.[8] All these statistics mean that the five million guestrooms must be cleaned, serviced, and maintained every day by one of the more than 1.9 million hotel property workers.

Think about stripping and making your bed with clean sheets every day. Then clean your toilet, bathtub, and bathroom surfaces, and dust, vacuum, and throw out the garbage from the day before. You will get tired just thinking about all these chores and instead decide to go to the gym. Not so for nearly four hundred thousand hotel housekeepers, the majority of whom are women of color and immigrants. These women must clean at least fifteen rooms or more each day, and at Hyatt up to thirty rooms. Some of the major corporate hotel chains, including Hyatt, Hilton, Starwood, and Marriott, have increased the speed by which the rooms must be cleaned, and the number of rooms each housekeeper is responsible for.[9]

Cleaning fifteen to thirty rooms a day is by any measure backbreaking work. But as a result of the "battle of the beds" started in 1999 by Westin, many of the beds changed daily are not your average hotel bed. The nation's premier hotels offer thick mattresses that would make the *Princess and the Pea* green with envy. Added to this are luxurious duvets, decorative bed skirts, and as many as six pillows gracing a king bed. Marketed to hotel guests as "Suite Dreams" and "Heavenly Beds," to the housekeepers they are anything but dreamy and heavenly.

A 2010 study of hotel worker injuries from fifty hotels published in the *American Journal of Industrial Medicine* found that housekeepers have a 7.9 percent injury rate, 50 percent higher than for all other hotel workers, and twice that for all workers in the United States.[10] Making matters worse, many of the hotel chains refuse to use fitted sheets, which would save the housekeepers from having to lift dozens of heavy mattresses each and every day. Other commonsense solutions have also been ignored, such as long-handled mops and dusters to reach high and out-of-the-way places.

Although the bed wars have been very profitable for the hotel industry, and comfortable for their guests, the injury rate for housekeepers has climbed to 71 percent more than for all hotel workers, compared with 47 percent more before 2002. A housekeeper who is assigned as many as thirty rooms a day at a Hyatt Hotel has as little as fifteen minutes to clean the room; change the linens and make the beds; scrub the toilet bowl, bathtub, and all the bathrooms surfaces; and dust, vacuum, empty the trash, clean the coffeepot, wash the mirrors, clean the hair dryers, and restock shampoo and soap. More than 75 percent of housekeepers surveyed report that workplace pain interferes with their daily routine activities, and two-thirds take pain medication just to finish their work assignments.[11]

In April 2012, OSHA reported the results of its investigation in response to complaints related to "ergonomic risk factors" at Hyatt hotels. Although OSHA did not issue Hyatt a general duty citation for these hazards, the findings identified "tasks presenting risk factors" for injury, including "repeated heavy lifting and carrying, bending, twisting, elevated and extended reaches, pushing and pulling, and forceful gripping."[12] OSHA recommended simple "intervention strategies," including long-handled tools for dusting, knee pads, fitted sheets, and lighter-weight vacuums and supply carts.

So, the next time you stay at a fancy hotel and sleep in a plush bed, don't forget the backbreaking work it took to provide you with a good night's rest. And don't forget to pick up your dirty towels from the bathroom floor and place them on top of the toilet seat or vanity counter instead. This little courtesy may save a housekeeper from bending down and injuring her back. And, finally, leave a generous tip on your way out the door.

Angela Martinez

Angela Martinez's body is breaking down after twenty-six years of work as a housekeeper for the Hyatt Regency Hotel in Chicago. At only four feet eight inches tall, Angela is nearly a foot and a half shorter than the Hyatt Grand Bed mattress that she lifts every day and which weighs in at just under one hundred pounds. Hyatt's doctors tell Angela that she is just getting old and that she has arthritis. Angela is getting old, but she knows better what is ailing her.

Iguala de la Independencia, in the state of Guerrero, sits in a valley halfway between Mexico City and Acapulco. It is a long way from Chicago and the famous Magnificent Mile near where the Hyatt Regency sits beside the Chicago River. Iguala is a historic city of around 110,000 persons, but with little industry save for its silver artisans. Tourists pass by and over on their way to the famous beaches of Zihuatanejo, Acapulco, and Ixtapa. When she was eighteen, Angela left Iguala for Acapulco, but not for the fun and sun. She went to study at the university, hoping to become an obstetrician gynecologist. Toward the end of her studies, she fell in love, married, and got pregnant with the first of her two children. She knew Rogacino Garcia from grammar school back in Iguala. He had just returned from living in Chicago, where his parents had moved years earlier. Rogacino graduated from high school in Chicago, and he too wanted to become a doctor.

After a while, Rogacino went back to Chicago, while Angela remained in Iguala, pregnant with their second child, Manuel. With two infants, medical school and becoming a doctor were now only a dream. Back in Chicago, Rogacino worked at different jobs, including in a juice factory. Eventually, in 1985, he was able to send enough money back for Angela and the babies to travel to Chicago. Angela was only twenty-five years old.

In Chicago, they all lived with Rogacino's parents, along with his sister and her family. Angela stayed at home with the children and made a little money taking care of neighbors' kids. She also took care of her sister-in-law's kids but wasn't paid anything for it. Hostilities brewed, and she got in a fight with Rogacino's sister, who accused Angela of stealing from her. Angela then decided it was time for her to get a real job. She received a work permit and applied to work at Hyatt in 1988. One of her friends worked there and told Angela it was a good place. She had never worked at a hotel before, but said she had on her application. She was tired of being humiliated by staying at home. Her marriage was also falling apart, and she would later separate from Rogacino. She was fighting with everything in her life. Angela wanted to reclaim some of her dignity through hard work.

The Hyatt Regency Chicago is an enormous hotel with two towers and more than two thousand guest rooms. It didn't matter what her level of education was; the only job available for her was as a housekeeper. On her first day of orientation, Angela learned how to make a hotel bed. She was taught to bend her knees when moving the bed to vacuum underneath. In 1988, making a hotel bed was a lot easier because the mattresses were smooth and not as large as the mattresses today in luxury hotels like Hyatt. Mattresses now weigh twice as much as they did when she first started with Hyatt. Major hotel chains are in bed wars over who has the most comfortable mattress, and thus the more mattress stuffing. The Hyatt Grand Bed website brags that their bed is "configured to give extra lower back support" for the guest. But not so much for the housekeeper. Housekeepers are the front-line casualties in the bed wars.

At Hyatt, Angela is assigned to clean sixteen rooms. She is lucky, because at some hotels housekeepers are assigned up to thirty rooms to clean. Still, she has only thirty minutes for each room. The rooms at the Hyatt are an assortment of standards and suites, with a king or two queen beds. Often there is also a roll-away bed. On an average day, Angela has to strip and make more than twenty beds. This involves repetitive pulling and tugging at bottom sheets, top sheets, pillowcases, and duvets to get them off and back on. All this tugging and pulling puts enormous stress on her hands, arms, and shoulders. Because Hyatt and other hotel chains do not use fitted sheets, Angela also has to lift the luxury mattress several times just to make a single bed. Imagine a gym workout regime of 25 "reps" of lifting a 100-pound weight five times. That's 125 "reps" lifting a total of 12,500 pounds. And that's just making the bed. There is lots more dangerous work to do in a hotel room. And there is no steam and juice bar at the end of this workout.

There are sixteen bathrooms to clean. Again, back in the day, Angela said, cleaning bathrooms with smooth ceramic tile was a breeze compared to the materials used now, porous cementlike surfaces that require more scrubbing and, in turn, more stress on her hands, wrists, arms, and shoulders. Her supervisor uses a cotton swab to inspect the tiniest nooks and crannies, so Angela has to get down on her knees to clean the tub, shower, and bathroom floor. And replacing towels that are left on the floors by unknowing guests (I used to be one) requires lots of bending down. Now, it is not enough just to neatly hang fresh towels on the rack. Angela has to roll thick towels in tight, neat rows. Her supervisor has asked Angela why some towels are rolled fatter than others. Angela told her, "Because they're like people. Some are fat. Some are thin," she recalled with a tired smile.

Out in the guest room, Angela has to pick up and clean the mess left behind by the guests. Most guests are neat, but many do things in and to the rooms that they wouldn't do at home. Guests with kids are the worst, she said. Kids leave messy handprints all over the furniture and windows. Cookies, Cheerios, and pretzels are ground into the carpet. Puddles of sticky, dried-up juice glaze the furniture. These rooms take the most time to clean.

Dusting the furniture, fixtures, and walls and vacuuming the carpet are every-room chores. Many housekeepers are not supplied with long-handled dusters, so they have to stand on chairs, sofas, and bathtubs to get to the hard-to-reach areas, and risk falls and other injuries. The vacuums are heavy and don't roll easily.

Finally, there is the supply cart. The cart, Angela said, is so heavy, loaded down with towels, sheets, mops, brooms, vacuums, and other supplies, that the wheels bend under the weight. When the cart is fully loaded, at around three hundred pounds, she has trouble seeing over the top of it as she pushes it down the carpeted hallway, which puts further stress on her legs, back, and arms.

All this would be manageable, except for one tiny detail. When she first started, Angela had fewer rooms to clean than she does now. She worked 8 a.m. to 4 p.m., with a thirty-minute lunch. Now she works until 4:30 and has a hard time cleaning all the rooms assigned to her. Angela said that in the rush to finish their rooms, housekeepers don't have the time to work safely. Housekeepers in a hurry to change a bed lift up the heavy mattress without first taking the time to slowly bend at their knees. Back and shoulder injuries among housekeepers are epidemic. Angela remembered one housekeeper who, working in a hurry, got tangled up in the sheets and fell to the floor, snapping her ankle.

In around 2002, the workers at the Hyatt voted for a union, UNITE HERE. Angela said that the union has helped bring the workers a voice in their workplace. The union negotiated additional breaks, which, like her lunch break, Angela often skips because she would not have enough time to finish her rooms. Still, she is happy to have a union, and has become a key leader at the hotel. Workers come to her with their workplace issues. "It makes me happy when I can help other workers," she said.

The toll from all this backbreaking work has finally caught up with Angela. In February 2012, she was experiencing continuous pain in her left arm and numbness in her hand. According to the emergency room doctor, she herniated a disc. The Hyatt doctors countered that she is fine. Just getting old. They prescribed ibuprofen and ice. The ibuprofen hurts her stomach and masks the injury, making it worse, so she lives with the pain. Angela said that there is often a line of housekeepers at the drinking fountain before the start of their shifts. They are all in line to take their ibuprofen. The Hyatt doctors also prescribed Angela five days of physical therapy. That didn't help, because after the therapy she continued to do the same tasks at the same unsustainable pace. There are no light-duty jobs at Hyatt. If she complains, they will send her home, and workers' compensation pays only 65 percent of her wages.

In April 2012, OSHA issued a formal hazard alert letter telling Hyatt what Hyatt workers already knew, that its housekeepers faced ergonomic risks every day on the job. The letter noted that OSHA observed routine housekeeping tasks that presented "ergonomic risk factors." Because there are no ergonomic standards—they were repealed by the Bush administration in 2001—OSHA was limited to "findings" that stopped short of more serious citations. Among the suggested "intervention strategies" suggested by OSHA was to supply housekeepers with long-handled dusters, a step stool, and knee pads. OSHA also suggested that Hyatt should consult with its employees "on evaluations of potential risk factors and on interventional strategies." In response, Hyatt stated: "The health and well-being of our associates remain one of our top priorities and, as we always have, we will continue to work with our associates to ensure that we provide a safe, healthy workplace."

Angela now lives with her daughter and baby granddaughter. After work she is too exhausted to help cook or clean the house. Instead, she sits on the couch with the baby and tries to stay awake. On the day that we spoke, Angela had ten rooms to clean, all with double beds. The double beds are harder to make than the kings. "They want to kill me," she said.

Electrician

> When Edison . . . snatched up the spark of Prometheus in his little pear-shaped glass bulb, it means that fire had been discovered for the second time, that mankind had been delivered again from the curse of night.
>
> EMIL LUDWIG, GERMAN HISTORIAN

Benjamin Franklin is often credited with "inventing" electricity when he attached a key to a string at the end of a kite and flew it into an electrical storm. The kite story has been told and retold in schools for generations as the genesis of electricity. Of course, Franklin's kite experiment didn't invent electricity, simply because electricity is naturally occurring. Nor did he discover electricity. Franklin's experiment demonstrated that lightning and electricity were one and the same, a not insignificant revelation.

Americans everywhere plug into the electrical grid to light the dark, cool their homes, and power their lives. The demand for electricity in the digital age is high. It has been estimated that desktops, laptops, and cell phones use about 10 percent of the world's total electricity generation, or the equivalent of Germany and Japan combined. Still, we take it for granted that the magic wall outlet will always provide reliable electricity for all our needs, and on demand. That is, until there is a power outage, and then we are relegated to near Neanderthal status, anxiously waiting for power to be restored so we can reconnect with the world. Try hooking up your iPhone to fire. Because of this, we are more dependent than ever on electricity, yet rarely stop to think about what it takes to generate the "spark of Prometheus" and the dangers that American workers face daily to keep us plugged in.

Surely, since we were all children, we were taught not to stick our tiny fingers into empty sockets or wall outlets, or remove a stuck piece of toast from a toaster with a metal knife. But that's about as far as most of us got with respect to comprehending the dangers of direct access to electricity.

For workers providing that access in the generation, transmission, and distribution industries, most notably power line workers, OSHA has a long-established standard, known as the "269" standard.[13] This requires employers to develop appropriate hazard prevention and control methodologies designed to prevent workplace injuries and illnesses. Employers are required to implement safe work practices and worker training. These include lock-out / tag-out procedures for deenergizing equipment, systems, circuits, and power lines, and then grounding the systems in the event that they become energized. Workers are also required to handle energized systems. In these instances, 269 requires, among other things, that the live systems be guarded or covered with insulating protective equipment, such as insulated line hoses, blankets, hoods, and barriers between the worker and the live system. In addition, it has been mandated that employers protect their workers by providing personal protective equipment, typically inexpensive insulated gloves and arm sleeves.

Still, electrical workers are injured, often fatally. The Electrical Safety Foundation International (ESFI), a nonprofit organization dedicated to promoting and providing education about electrical safety practices at home and at work, reported in a study of electrical fatalities from 1992 to 2010 that 44 percent of workplace electrical fatalities were from working on or near overhead power lines, while 27 percent of fatalities were from contact with transformers, wiring, or other electrical components. According to ESFI, in the seven-year period from 2003 to 2010, there were 1,738 electrical fatalities, or 248 every year.[14]

Unprotected contact with electricity is often fatal. The human body is a perfect conductor of electricity. An electrical current passing through a worker's body will cause skin burns, damage to internal organs and other soft tissues, cardiac arrhythmias, and respiratory arrest. An electrical current traveling between arm and arm, or between an arm and foot, is likely to traverse the heart, causing arrhythmia. Low-voltage 60-Hz alternating current is the standard in residential homes for powering everyday appliances. However, a 60-Hz current traveling through the chest for even a fraction of a second can cause ventricular fibrillation. At the very least, contact from a low-strength electrical field causes an immediate and un-pleasant feeling of being "shocked." Contact with a high-strength electrical field that electrical workers are more likely to be exposed to will cause thermal or electrochemical damage to internal tissues. Damage will likely include (in clinical terms) massive edema, hemolysis, protein coagulation, coagulation necrosis of muscle and other tissues, thrombosis, dehydration, and the tearing away of muscle and tendon. Finally, an electrical shock can actually propel a worker off a roof or ladder.

In 2015, BLS reported that there were 425,070 workers classified as electricians in the construction industry, a designation that the slightly more than 14 percent of electricians who were represented by a union would take issue with.[15] OSHA defines "qualified workers" under the electrical safety regulations as those specially trained and certified to work on electrical equipment.[16] A self-described experienced electrician is not "qualified" by OSHA unless the employer can show proof of the appropriate training and certifications. Training has historically come through lengthy programs jointly sponsored and taught by the International Brotherhood of Electrical Workers (IBEW) and the National Electrical Contractors Association (NECA). IBEW-NECA training programs typically last four years and involve extensive classroom and on-the-job training with certified journeymen. The National Fire Protection Agency (NFPA) is a leading organization on developing standards, codes, and training on eliminating dangers due to fire, electrical, and related hazards. According to NFPA's 70E safety standard, 67 percent of work-related electrical injuries are attributable to unsafe acts.[17] This means that training and employing "qualified workers" may be the most effective antidote to electrical injuries and fatalities in the workplace. But "qualified" electricians are more expensive, usually because they are represented by a union that negotiated higher wages than those for nonunion workers. In addition, union workplaces allow only "qualified" workers and have negotiated safety committees and rules, and as a result are safer. However, construction and maintenance firms looking to add to their bottom lines compromise safety by employing unqualified workers. Safety training makes a life-and-death difference.

If you are reading this on a Kindle, or other similar device, under the soft light of an energy-efficient lightbulb, while listening to iTunes and drinking your home-brewed tea or a cool glass of white wine, remember to thank not only Thomas Edison, but most of all, the workers who keep you reliably connected.

Paul King

Medford, Massachusetts, sits astride the Mystic River just a few miles from downtown Boston. It was settled in 1630 and was home to the private plantation of Governor Matthew Cradock. By the end of the seventeenth century there were around two hundred people living in Medford. Tufts University was chartered in Medford in 1852. In 1900, there were more than eighteen thousand persons in Medford, drawn there over the years by industry manufacturing brick, tile, rum, clipper ships, and something called "Medford crackers." The 2010 U.S. Census reported 56,173 souls in Medford, down from its peak of 65,000 in 1960, but on the uptick, largely owing to young families moving there for its schools and proximity to downtown Boston.

Medford is predominantly white, Irish, and Catholic. Hardworking middle-class families send their kids to Christopher Columbus Elementary School or one of the three parochial schools. There is also Medford Vocational Technical High School, or "Voc." That's where Paul King went to school to learn a trade.

His mother planned to name him Robert. Instead, he was born on Paul Revere Day in 1955. Medford is one of the towns that his namesake rode through on his "midnight ride." Paul was one of six children—five boys and a girl. His sister, Mary Ellen, never married and worked for the phone company. She died of breast cancer in January 2003. The brothers took her death very hard. They all lived around the Medford area and were a close family. They helped fix one another's houses. Every summer, and on the holidays, they went up to Maine with their families, including eighteen cousins. The family photo albums look like every other family album. But on closer inspection, it is apparent that Paul is the center: resting on the living room couch with one of the infant girls lying on his chest; at the shore and in the water with the kids; celebrating life cycles and anniversaries; on cruise boats and vacations. Always, there are lots of people and lots of activity in the still photos. His life was jam-packed with his family and his work.

At Medford Vocational, Paul wanted to learn how to be an electrician, but those spots were all taken. So he learned how to be a printer. His older brother was a printer. While there, he met Debbie Green. She was at Medford High School. They dated for six years before marrying in 1978. Paul and Debbie had three children: Gina, Melissa, and Michael.

The first printing press in North America came to Cambridge, Massachusetts, two years after Harvard College was founded in 1636. It was brought over from England by the Reverend Joseph Glover with his other possessions and his wife, Elizabeth. As Harvard grew in prestige, Cambridge became the printing center in the American colonies. A second press arrived from England in 1659 and was set up in Harvard Yard. When a rival attempted to bring a press from England to Boston, Harvard used its considerable clout and successfully lobbied for a law prohibiting printing outside of Cambridge. The rival, Marmaduke Johnson, opened the first independent printing press in the colonies in Cambridge. Cambridge remained a major printing and publishing center into the twentieth century. Henry Houghton published Longfellow, Thoreau, Hawthorne, and Henry James. He eventually partnered with George Mifflin and founded Houghton Mifflin.

Paul graduated from Voc in 1973 and went to work as a printer. He started as a feeder, and then moved to a second man and finally to a pressman. He worked mostly for commercial printing companies. He liked printing. He liked to work with his hands. Paul liked to work. It was good work when he started, and he often worked seven days a week, getting double time on Saturday and Sunday. All three kids went to parochial school K–8, and then to nearby Arlington Catholic High School. After that, all three went to college. He paid all of their tuition. "He took care of his family," Debbie said in her strong Boston brogue. "That's all he knew all his life was hard work."

The glory years of commercial printing everywhere, including Cambridge and the Boston area, gave way to digital alternatives, such as online media. Companies shut down or consolidated. Pressmen were laid off. Around 2004, the industry contraction hit Paul's company, and it closed its doors. For the first time since 1973, Paul was out of work. He had no job and no prospects.

He took odd jobs to help pay the bills and to work and feel useful. He worked as a bartender. He worked for a catering company. He was a maintenance man. Debbie worked as a bookkeeper. She said it was tough on Paul to be out of work. In 2004 he went back to technical school to get trained as a HVAC technician. It was a twelve-week course to become certified in heating and air conditioning systems.

In late May 2005, he finally found a job. He was hired by a company called MainTech. It was a subcontractor for JetBlue Airways at Logan International Airport in Boston. JetBlue was moving into Terminal 3, and MainTech was hired to do maintenance work in the terminal. Paul was initially hired to do odd jobs like changing locks. He worked every day in June. Debbie said there were not enough maintenance workers. There was a separate subcontractor to do electrical work for JetBlue.

On July 27, 2005, Paul was directed by his supervisor to go up to the terminal roof and, with the help of another employee inside the terminal tech room, pull wires up to the roof to the compressor. The fan in the tech room in the terminal was broken. Massport, the public agency that manages Logan, said that there was a break in the line. MainTech was responsible for fixing the problem and replacing the wires.

Up on the roof, Paul was communicating by Nextel phone with his work partner down in the tech room. He was having difficulty pulling the wires up to the roof. According to the OSHA inspection report, the conduit wasn't big enough to pull the wires through. After five minutes passed without any more wire being pulled, his partner tried to raise Paul by the Nextel. He received no response. He tried a few more times and heard only beeping sounds. He then went outside and called up to Paul. Again, there was no response. He climbed up to the roof, where he found Paul lying on the ground. His head was facing away from the wall. His eyes were rolled back in his head. There was blood on the side near his eyes. And his glasses were on the ground, bent and smudged. The partner ran to the edge of the roof and yelled down to some JetBlue employees to call 911. Two of the JetBlue workers climbed to the roof. One of them began to give Paul CPR. After a few chest compressions, Paul started to spit out fluids through his mouth and nose. Paramedics arrived three minutes after they received the call. Paul had no pulse and was not breathing. The EMT attached a defibrillator and attempted to shock Paul back to life. The EMT reported that Paul had "thermal burns on both hands, face, shoulder and had blood stained glasses." Paul was removed from the roof and transported to Massachusetts General Hospital, where he was reported dead at 10:51 a.m. The autopsy reported deep electrical burns to both his hands.

The OSHA report surmised that Paul had bent at the waist, with his shoulder on a 440-volt electrical box in order to apply leverage and pull the wires through. He was a big guy, and he apparently thought he could muscle the wires up to the roof. However, Paul didn't know that the electrical box was energized when he leaned into it. A human body is a perfect conductor for 440 volts of electricity. He was immediately electrocuted.

Voltage is a measure of the fundamental force or pressure that causes electricity to flow through a conductor. Electrical injuries consist of four main types: electrocution, electric shock, burns, and falls caused as a result of contact with electrical energy. Electrocution results when a human is exposed to a lethal amount of electrical energy. For death to occur, the human body must become part of an active electrical circuit having a current capable of overstimulating the nervous system or causing damage to internal organs. Electrical injuries may occur in various ways: direct contact with electrical energy, injuries that occur when electricity arcs to a victim at ground potential (supplying an alternative path to ground), flash burns from the heat generated by an electrical arc, and flame burns from the ignition of clothing or other combustible, nonelectrical materials. Direct contact and arcing injuries produce similar effects. Burns at the point of contact with electrical energy can be caused by arcing to the skin, heating at the point of contact by a high-resistance contact, or higher voltage currents. Contact with a source of electrical energy can cause external as well as internal burns. Exposure to higher voltages will normally result in burns at the sites where the electrical current enters and exits the human body. High-voltage contact burns may display only small superficial injury; however, the danger exists of these deep burns destroying internal tissue.

According to a report from the National Institute for Occupational Safety and Health (NIOSH), the speed with which resuscitative measures are initiated has been found to be critical. Immediate defibrillation would be ideal; however, for victims of cardiopulmonary arrest, resuscitation has the greatest rate of success if CPR is initiated within four minutes and advanced cardiac life support is initiated within eight minutes.

Debbie got a call from a social worker at Mass General, who told her, "Your husband was seriously hurt." She was told her to come right away and "don't drive yourself." The caller did not report that Paul had already been pronounced dead. Debbie put down the phone and remembers that she couldn't call their kids because she had become hysterical. She remembers that she had a Nextel phone that communicated with Paul's brother, Bernard. Debbie did some work for Bernard. She reflexively beeped Bernard and told him, "911. Call the office." He called the work office and learned about an accident and that Debbie was on her way to Mass General. Bernard hung up and called Michael. He reported, "Dad's been taken to Mass General." Michael assumed it was their grandfather. Michael called his sisters to sound the alarm. They all headed to Mass General thinking something had happened to Grampy. On the way to the hospital, Michael called Debbie. He asked about his grandfather. She replied, "It's not your Grampy."

When Debbie arrived at Mass General, she was met by the social worker. They went into a little room and sat down. Debbie was told that Paul had died.

Melissa, their middle daughter, was in graduate school in Providence when she got the call to come to the hospital for Grampy. When she arrived at Mass General, Michael met her outside and told her that Dad had died. As for Grampy, he was too broken up to come to the hospital.

Paul's wake was on Saturday, and he was buried on Sunday. Debbie described the crowded scene at the church. "It was like Easter Sunday." Gina and Melissa eulogized their dad.

Melissa described Paul as "Mr. Fix It, or so he thought." She told a story of her dad repairing a high heel that she broke before going out for the night. He took the heel and shoe into the garage and banged on it for twenty minutes before returning it to Melissa. It resembled her shoe, so she wore it out that evening. "It didn't matter that my right leg was four inches taller than my left." She thanked him in the eulogy for giving them everything "we could ever want, and without him the three of us would be nowhere near where we are today."

Gina remembered Paul's hands. She told about how his hands held her as an infant, steadied her bike, and taught her how to swim. His hands cheered her achievements and worked day and night to provide for his family. Gina said, "To me your hands were your heart. I am honored and privileged to hold your hand and heart forever in mine."

Paul had just turned fifty the previous April 19. For his birthday, his kids took him skydiving. He was not into extreme sports or dangerous activities. But he took the challenge and jumped out of a plane by himself at ten thousand feet. He loved the jump and loved to retell the story. Mostly, he loved his kids and Debbie. He is now buried in Medford, where he grew up.

A supervisor for Massport reported that two licensed electricians should have been on the roof together doing the job that killed Paul King. Paul was not an electrician. After he was laid off, he talked about going back to Voc to finally get certified as an electrician. But he never did. Paul's partner was also new to MainTech, having been laid off as a custodian and cafeteria utility person from a nearby school only three weeks earlier. He was not a licensed electrician. He was still in training as a maintenance person at the JetBlue terminal. He reported to the state trooper investigating the accident, "I'm not an electrician, and I don't know whether Paul was one." After Paul's death, Massport conducted a months-long study that concluded that it needed to hire four additional journeymen electricians. Massachusetts law requires that two licensed electricians are required for working on equipment as powerful as 440 volts. Paul was alone on the roof, and he was the second worker electrocuted at Logan within a year. Debbie said that working on the roof, "Paul was in a bad situation, but he didn't know it."

The OSHA investigation revealed that the day before the accident, Paul's supervisor directed Paul and his helper to "snake" wires up to the roof. The first thing they did was to shut the power off and lock out the box so no one could turn the power on while they were working. They were the only ones who had keys to unlock the power source. The problem was that, being untrained, they shut off and locked the wrong power source. Metal pliers were found near Paul's body. OSHA concluded from the pliers and the burns on his hands that he was "holding the metal portion of the pliers with both hands while he was pulling the wires. The end of the wire made contact with the energized parts which shorted out against the nut connecting the box to the conduit."

OSHA issued MainTech eight citations for "serious" violations. The violations included not providing training to employees on lock-out / tag-out procedures for electrical power sources, or other safety practices related to electrical equipment. Significantly, MainTech was cited for not providing employees with PPE, personal protective equipment, such as safety glasses and dielectric gloves. The OSHA inspector concluded, "Had the employee been wearing protective gloves there would have been a chance that he would have survived." Rubber lineman gloves, with a rating of 500 volts, cost less than fifty dollars. Paul also did not have electrical test equipment with him on the roof that would have enabled him to test whether the electrical box was energized.

His supervisor told OSHA that he did not specifically assign the job of fixing the fan in the tech room to Paul, but that it "was supposed to be done in the future." He further told the OSHA inspector that he "believed" that Paul was an apprentice electrician and that he thought he was "qualified to run wires," but never saw any qualifications to support his belief. He added that Paul had done some electrical work for him in the past and that "he did a great job." The OSHA report found that neither Paul nor his helper was "trained on electrical hazards," and that they were not electricians or apprentices.

OSHA fined MainTech $54,000. MainTech's manager of employee satisfaction wrote to OSHA stating that it "intends to contest all of the items and penalties alleged in the Citation and Proposed Penalty." The challenge to the fines was not filed in a timely manner. OSHA eventually collected the penalties, but it took awhile before MainTech paid.

Paul's final paycheck from MainTech did not pay him for the time he worked on the day he died. Melissa complained, and a check was eventually sent to Debbie for the time he worked. A couple of days after Paul's death, Debbie received a $5,000 workers' compensation death benefit. Later she began receiving weekly benefits from workers' compensation. "I still receive weekly benefits, unless I get remarried or win the lottery," she said with a wry smile. State investigators from workers' compensation come around from time to time to check up on Debbie to make sure that she hasn't remarried or come into a fortune.

As is the case in all states, Debbie couldn't sue MainTech for Paul's death because he was killed at work. Workers' compensation shielded MainTech from any claims of negligence for his death. Nonetheless, the kids and his brothers insisted that Debbie hire a lawyer. She sued JetBlue and Triangle, MainTech's parent company, for wrongful death.

The best day for a plaintiff is the day the lawsuit is filed. After that, litigation is at best a slog and most often painful; it doesn't close the open wounds fast enough. True to form, the litigation lasted for five years and was difficult for Debbie and the family. The companies asserted that Paul was at fault. They claimed that he broke the law by doing the work without being a licensed electrician. "We felt like we were on trial," Debbie said. "We had to prove that Paul was a good guy. A good provider and father." In the end, they settled in the middle of the trial. The end of litigation is frequently inconclusive and unsatisfying. It was for Debbie. Paul is still gone, and her grief remains forever.

Gina and Melissa got involved in the Massachusetts Coalition for Occupational Safety and Health. MassCOSH advocates on behalf of worker safety. Melissa is on the MassCOSH board of directors. She has lobbied to increase funeral benefits for workers killed on the job. Gina and Melissa got Debbie to volunteer as a bookkeeper for MassCOSH. Debbie contributed some of the settlement money to the United Support and Memorial for Workplace Fatalities, a national support group for survivor family members.

Gina got married after Paul died. The wedding was bittersweet. She has two young boys and now lives in Minneapolis. Melissa works in marketing and lives near Debbie. According to Debbie, Michael struggled with Paul's death. "He lost his direction" without his father. He took time off from college and had a hard time concentrating. They were close. Paul coached Michael's hockey teams and went to all his games.

Debbie was in a bad way for a long time after Paul's death. Even today, little things will bring up tears. "We try and tuck it down," she said. "But the grief still comes up."

Coal Miner

> Where the sun comes up about ten in the morning
> And the sun goes down about three in the day
> And you fill your cup with whatever bitter brew you're drinking
> And you spend your life digging coal from the bottom of your grave
>
> DARRELL SCOTT, "YOU'LL NEVER LEAVE HARLAN ALIVE"

The "Timeline of Coal in the United States," published by the American Coal Foundation, a provider of coal-related educational materials founded and supported by coal producers, dates the use of coal in the Americas to 1000 CE, by the Hopi Indians. The Hopi used coal to bake pottery made from clay. The timeline has twenty-six entries, including such seminal events as Baltimore becoming the first city to light its streets with gas made from coal, and the invention of the coal shuttle car in 1937.[18] However, missing from the timeline is any mention whatsoever of the human toll that coal mining has exacted on hundreds of thousands of men and women.

From 1900 to 2015, "mine disasters"—officially defined as an accident where five or more miners are killed—took the lives of 11,719 coal mine workers.[19] Most of the disasters were from methane explosions. Until the 1980s, the canary in the coal mine was still used to detect the tasteless and odorless deadly gas.

In the timeline of coal-mining deaths in the United States, 1907 was the deadliest year, when an estimated eighteen mine disasters killed 3,242 miners from explosions or other accidents. The single deadliest mine disaster still belongs to 1907, when 362 miners were killed in the Monongah, West Virginia, explosion.[20] This led to the establishment in 1910 of the United States Bureau of Mines. Dead coal miners have historically correlated with the passage of mine safety laws. In 1968, an explosion at the Farmington No. 9 Mine in West Virginia caused the death of seventy-eight miners. The Federal Coal Mine Health and Safety Act of 1969, creating the Mining Enforcement and Safety Administration, soon followed. In 1976, twenty-six miners and their rescuers were killed at the Scotia Mine in Kentucky. The Federal Mine Safety and Health Act of 1977 strengthened the 1969 act and relocated the agency to the U.S. Department of Labor. Finally, 2006 saw seventy-three coal miners lose their lives, including twelve miners at the Sago Mine in West Virginia who died of carbon monoxide poisoning after a methane gas explosion. The dead miners were found seven hundred feet from fresh air. The 2006 MINER Act resulted, requiring, among other things, that mine operators stow along escape routes caches of self-contained breathing apparatus that must supply at least two hours of oxygen per miner.[21]

Although deaths and mine explosions trapping miners receive the biggest headlines, the reported number of injured miners is annually in the thousands. Contact with objects, debris, and equipment, and transportation accidents, are common causes of mining injuries. In a Bureau of Labor Standards report, fatal injuries to miners in 2007 were six times that of private industry, while the rate of nonfatal injuries and illnesses in bituminous underground coal mining in 2008 was 66 percent higher than that of all private industry. The incidence of illnesses and injuries requiring days away from work in coal mining was twice that in the private sector as a whole.[22]

From the moment a miner enters a coal mine, he is at higher risk to be killed or to incur a severe nonfatal injury than any other worker. In a process called "retreat mining," a coal miner's workplace is not big enough to stand in, and it is prone to roof collapses as the cutting machine advances along the coal seam. The most common method of underground mining is the room-and-pillar method where coal seams are mined by a machine known as a "continuous miner" that cuts a grid of "rooms" into the seam. As the rooms are cut, the continuous miner simultaneously loads the coal onto a shuttle or ram car, from which it will eventually be placed on a conveyor belt that will move the coal to the surface. "Pillars" of coal are left behind to support the roof of the mine. As mining continues, roof bolts are placed in the ceiling for support and to avoid ceiling collapse.

Once the end of the seam or property line has been reached, "retreat mining" begins. The pillars of coal holding up the roof, like legs on a table, are cut away, causing an intentional roof collapse. Because of this, retreat mining is the most dangerous type of mining there is, as if the insides of the mountain have been carved into Swiss cheese. Once the last bit of coal has been removed, the mine is abandoned, leaving a hollowed-out mountain that has caved in on itself, often burying with it the bodies of unlucky miners caught in a roof collapse.

If coal miners are lucky enough to avoid death and injury from a mine disaster or other accident, they are likely, by virtue of how long they have survived, to get sick and die a painful death from black lung disease. Black lung, also known as coal workers' pneumoconiosis, is caused by the prolonged inhaling of coal mine dust, mostly found in underground mines. It scars the lungs, making healthy tissue look like something that was charred over a hot flame. Emphysema, shortness of breath, airway obstruction, and premature death all follow.

From 1968 through 2007, black lung caused or contributed to more than seventy-six thousand deaths in the United States, costing $45 billion in benefits to miners and their families.[23] In 1968, miners went on strike in West Virginia, demanding that the legislature pass a bill compensating miners afflicted with black lung. The governor eventually relented and signed the bill. The next year, Congress, in the Federal Coal Mine Health and Safety Act, mandated the eradication of black lung. The law provides monthly benefit payments to coal miners totally disabled as a result of black lung, or to the widows and dependents of coal miners who died from the disease.[24] In 2016, a miner or his widow received $644.50 a month, offset by any other benefits from workers' compensation, unemployment compensation, or disability insurance payments. For a widow and three or more dependents, the family benefit in 2016 for a miner who died from black lung paid $1,289 a month, or $15,479 a year.[25] The federal poverty level for a family of four in 2016 was $24,250 a year.[26]

By the late 1990s, cases of black lung disease had declined to 2 percent of miners screened, down from 11.2 percent when the 1969 legislation was passed. The decline in black lung rates was largely attributed to regulations to reduce miners' exposure to respirable coal dust through ventilation and suppression of dust.[27]

Then, in 2005, screenings began to show a resurgence of black lung, especially in central Appalachia. In Eastern Kentucky, 9 percent of miners screened between 2005 and 2009 had black lung. More alarming, the disease was occurring in younger miners, and with severe cases of fast-progressing black lung on the rise. Of the twenty-four miners who died in the 2010 Upper Big Branch disaster, and who had enough lung tissue to autopsy, seventeen had signs of black lung, with some of the miners having worked less than ten years in the mine. Government and industry experts have identified a number of reasons for the resurgence in black lung, including routinely working ten to twelve hours a day; inadequate dust control rules; and the failure by coal producers to comply with the rules. Cheating on dust samples submitted to federal regulators is believed to be a significant contributor. New rules and regulations to protect miners were delayed by the coal industry's opposition.[28] Too late for many miners, on August 1, 2014, a new respirable coal dust rule went into effect to better protect miners from coal dust.[29]

In the end, miners are still dying from accidents, while black lung leaves miners' lungs scarred, shriveled, and black as they struggle to breathe while the screw from the disease is slowly tightened across their throats.

Scott Howard

Charles Scott Howard's blood must run black like the coal he has mined for more than thirty years. Before him, his grandfather, or "pappo," started working in the mines in 1917, living in company housing called camp houses. He fought to bring in the union in those turbulent years in the 1920s and 1930s. His pappo was a master carpenter for Peabody Coal. He worked until the mid-1960s, including many years slowly choking to death from black lung. Like his pappo, Scott has been diagnosed with early stages of black lung, a disease that was once believed eradicated.

Scott is from Evart, Kentucky, in Harlan County. Living there, in the small towns or hollows tucked into the mountains of eastern Kentucky, a young man either went into the military or down into the mines. Since 1911, when commercial coal first was shipped by rail out of Harlan County, coal mining replaced subsistence farming as the dominant industry. Scott initially followed in the footsteps of his father, a career army man. However, after a short stint in the army, Scott went underground at age nineteen, returning home to take care of his baby sister who was brain damaged. "My family needed me more than the military," he said.

Scott started at the Tracer Coal Company. It was a small mine, and he was a "jack setter," the miner who places hydraulic jacks, timbers, and posts in specified locations, usually every four feet, to form a roof support system in the mine. He called the supports "ignorant sticks," because miners who set the supports risked getting chewed up by the augers from the continuous miner machine that cuts into the coal seam. The workspace for a jack setter, and for the other miners involved in cutting the coal out of the mountain, is by any measurement a confined space. The average height of the space is three to four feet, never high enough to stand up in. Scott and his fellow miners work on their knees, crawling around in the dark for twelve hours a day, illuminated only by the lights from their helmets and the headlights of the mining machine.

Scott lasted around a year at Tracer Coal before it went out of business. He recalls that it ran out of money to pay for the electrical generators it leased. Scott claims that he has worked at more than forty mines over his career. Many mines would close up just to get rid of the union, and reopen nonunion some time later. The coal wasn't going anywhere, Scott explained, so this tactic didn't really cost the coal companies. If anything, he said, it made the operators more money, since they would eventually reopen and continue extracting the last bit of coal, this time with miners being paid less than when the union was there. Today there are no more union miners in eastern Kentucky.

When it comes to the treatment of coal miners and their families by the coal producers, there is no doubt which side Scott is on. Not only do

the companies extract every chunk of coal they can, leaving the moun-
tainsides sagging and collapsing; they take all their profits with them.
"All the money goes to Frankfort and Lexington," where the bosses and
politicians live. Their schools are built and roads paved with coal money.
Nothing is sent back by the politicians to the coal counties like Harlan.
"Coal operators have always only wanted slave labor, where the men have
to come to the operators for a job and their lives." Scott added, "There is
nothing here for a young person except to get in trouble."

Early in his mining career he didn't know what was legal or not in
terms of mine safety. Scott used to set wooden supports to hold up a ceiling from caving in. Now he knows that hydraulic supports, or automated
temporary roof supports (ATRS), are required while the "roof bolters" in-
sert metal rods into the mine roof from their knees. Instead of supporting
the roof with wooden timbers from beneath, which had been the practice
until the late 1940s, roof bolts are inserted directly into the rock to make
the roof self-supportive. Although this has made the mine safer from roof
falls, the job of the roof bolter has remained one of the most dangerous
in mining.

Scott was a roof bolter at a mine in the late 1980s. He was fired for
"insubordination," because he refused to bolt a roof without an ATRS to
hold up the roof. With the union gone from Appalachia, miners had few
options for help in redressing their safety complaints and mistreatment.
When he got fired, Scott didn't go away quietly to find work in the next
mine. Instead, he found Tony Oppegard, a Lexington lawyer and mine
workers' advocate. Together, they made a formidable team, the coal operators' worst nightmare.

From the earliest days, coal operators relied on poverty and fear to
keep miners down. With the union driven out, they had no real oppo-
nent to their goal of extracting coal while maximizing their profits, often
by putting profits ahead of safety. But Scott, backed by Tony Oppegard
and the Appalachian Citizens' Law Center, decided it was more important
to fight for mine safety, and fight like hell he did. Oppegard filed a discrimination complaint with the Mine Safety and Health Administration
(MSHA), claiming that Scott was terminated for refusing to work under
unsafe conditions. MSHA wanted him to settle for $1,500 and to give up
his job. Scott and Oppegard said no. In the end, Scott settled on favorable
terms but found himself blackballed in Harlan County as a troublemaker.

He was forced to work in other counties, but continued his fight for mine safety. Scott Howard and Tony Oppegard together have filed more than a dozen complaints with MSHA against the coal operators. "When I saw something wrong, I'd tell the boss," Scott said. "I've put one foot on a banana peel and my neck in a noose, but it's worth it when you do the right thing," he said. Oppegard is a little less slapstick in describing Scott Howard's courage. "Howard has paid a toll for all this. Your boss doesn't like you. It's unsafe. You have to look over your shoulder."

Scott's biggest battle was with the Cumberland River Coal Company, a mine owned by Arch Coal, one of the world's largest coal producers. In July 2007, Scott testified before MSHA at a hearing in Lexington. The subject of the hearing was faulty mine seals, which were cited as the cause of two mine explosions in 2006, the Darby and Sago mine disasters that killed a total of seventeen miners. Mine seals are constructed to keep atmospheric pressures separate in the mines in the event of an explosion, such as from methane. The seals are measured in pounds per square inch, or psi, and before the Darby/Sago disasters, the seals were designed to withstand only 20 psi. At the hearing, Scott showed a video of seals in an unnamed mine, although all in attendance, including MSHA inspectors, knew it was from the Cumberland River mine where he worked. The video showed fractured seals, some with water coming through the broken seals. Before Scott had traveled the 150 miles back home, MSHA inspectors were already at the mine inspecting the broken seals. Cumberland River was cited for safety violations. Scott Howard was a marked man at Cumberland.

Less than two years later, he was laid off at Cumberland as part of a mass layoff, even though he was the most senior employee to get laid off. He filed a complaint with MSHA that his layoff was a result of his safety activities, and an administrative law judge agreed and temporarily reinstated him. Once reinstated, Scott took on the issue of ventilation problems in the mine. A properly ventilated mine brings in fresh air to dilute and offset the toxic air from mine dust and gases. Miners exposed to respirable coal dust are likely to develop black lung. When his complaints went unheeded, Scott literally blocked the roadway in the mine that brings coal to the conveyor belts. He said, "You give me air, and I'll give you coal." He got clean air.

Scott wasn't disciplined for this. But something worse happened. On July 26, 2010, he was working alone in an area of the mine moving mine waste. He has no recollection of what occurred next, except from the MSHA and hospital reports that followed. Something, or someone, struck him on his head with such force that it shattered his helmet and broke his teeth. He was knocked unconscious, and when he came to he was dazed and confused, driving his mine car like he was drunk. Some miners found him and got him out of the mine and to a hospital. His diagnosis was traumatic brain injury. For ten months he was off work and in rehab. He suffered from cognitive problems, double vision, memory lapse, and even a personality change at times. He was seen by seven different specialists in addition to his primary care doctor.

A couple of days before the accident, Scott had applied to MSHA to become a "miner's rep," a miner who accompanies MSHA on its mine inspections. Scott doesn't believe in coincidence. He thinks someone tried to kill him. The official company line on the accident is that maybe a water line broke, causing a rock to hit Scott in the head, although there was no evidence of a water line break.

After ten months, Scott was released to return to work. Seven of eight physicians he saw released him to work without restrictions. The eighth, Dr. Robert Granacher, released him, but with restrictions "not to work at height." His supervisor stated that Scott could be accommodated for this one restriction. However, Cumberland River management asked Dr. Granacher for clarification as to the meaning of "at height." In the meantime, Scott returned to work and was undergoing an annual retraining. On the day of his retraining, Cumberland received Dr. Granacher's clarification. He now opined that Scott could not work at any job and should be permanently restricted from underground mining. Based on this, Cumberland told Scott to leave the retraining and fired him. His workers' compensation benefits were also terminated.

That same day, Scott and Oppegard filed a complaint with MSHA alleging that he was prevented from returning to work because of previously engaging in safety activities. An administrative law judge (ALJ) ruled that Cumberland had unlawfully terminated him and ordered him reinstated. Cumberland appealed the decision to the Sixth Circuit Court of Appeals, and on April 4, 2013, the court affirmed the ALJ. The ALJ and the Court of Appeals wrote that Scott had engaged in significant protected activity by filing seven complaints against Cumberland. They also rejected Cumberland's claim that, based on Dr. Granacher's clarification, Scott could not safely work in the mine. Significantly, the ALJ and the court focused instead on e-mails and other communications among Cumberland management that expressed their intention and plan to keep Scott from working. The e-mails went so far as to suggest that Dr. Granacher's restrictions might force a settlement, instead of Scott coming back to work "in the driver's seat (not what we want)." In the end, the court rejected Cumberland's "purported safety argument that is supported by only one physician's opinion," while ignoring the opinions of seven other doctors. The court concluded that Cumberland's "justification was simply a pretext designed to mask the true reason for Howard's termination"—his safety advocacy on behalf of miners.

When Scott Howard returned to work, he returned with a federal MSHA inspector by his side. He was a "miner's rep," and he was more determined than ever to fight for a safe and healthy workplace. Scott said about his return, "You would have thought I owned the place."

Oil and Gas Worker

America is addicted to oil.
PRESIDENT GEORGE W. BUSH, 2006 STATE OF THE UNION

President Bush's declaration was more noteworthy than newsworthy because it came from the son of a family of wildcatters. It was less newsworthy because America's addiction to oil has been widely known and discussed for years. Yet despite a measurable transition to clean energy, with consumption from traditional energy sources down or flat and renewable energy rising, Americans are still addicted to oil.

Fueled by low gas prices, in 2015 Americans guzzled 9.16 million barrels of gasoline *per day*, or 140.43 billion gallons, the U.S. Energy Information Administration (EIA) reported.[30] But gasoline is not the only reason for our petroleum addiction. Everyday Americans use petroleum-based consumer products, including detergent, plastic, vitamins, CD/DVDs, synthetic fibers, fertilizer, candles, and Band-Aids. To serve our addiction, the United States replaced Russia as the top oil and natural gas producer in the world, a changing of the guard with wide-ranging economic, foreign, and security policy implications.[31] And, to ensure a steady flow of oil, new technologies have permitted extractions from previously untapped sources. Oil is now produced from sands and shale.

Of the primary energy consumption by sources in the United States in 2015, petroleum was the largest at 35 percent, with natural gas a close second at 28 percent.[32] Natural gas is produced from shale by the controversial extraction method of "fracking," or hydraulic fracturing. In the last ten years the demand for natural gas has exploded. Natural gas is mostly used for heating, but is also used as a raw material for many products, including paints, fertilizer, plastics, antifreeze, dyes, photographic film, medicines, and explosives.

The oil and gas industry's trade association, the American Petroleum Industry, reports that 9.8 million jobs are supported across all segments of the industry, including exploration, production, processing, transportation, and marketing, and make up 8 percent of the U.S. economy.[33] Exploration and extraction occur both offshore and on dry land. Transportation includes the millions of miles of pipelines crisscrossing the United States, as well as transportation by land and sea. Within these broad segments, there are hundreds of specifically defined job classifications. In extraction and processing, they include operators, roustabouts, and pumpers working at the wellheads and refineries.

From 2003 to 2010, OSHA reported that 823 oil and gas extraction workers were killed on the job, a fatality rate seven times greater than the rate for all U.S. industries. Safety and health hazards in extraction come from vehicle accidents, explosions and fires, falls, confined spaces, high-pressure lines and equipment, ergonomic hazards, electrical and machine hazards, and chemical exposures.[34] In some disasters, all these hazards have occurred at one time—when, for example, eleven workers were blown off the oil platform in the Gulf of Mexico in the BP/Transocean Deepwater Horizon disaster.

From 2003 to 2013, the U.S. oil and gas industry experienced unprecedented growth financially and in the sheer number of its operations. During this boom, the workforce in the extraction industry doubled, and the number of drilling rigs increased by 71 percent. Workplace fatalities in oil and gas extraction during this ten-year period increased by 27.6 percent, for a total of 1,189 deaths. According to a study by the Centers for Disease Control and Prevention (CDC), for this period the annual fatality rate resulted on average in 108 deaths per year, and an incidence rate of

25.0 deaths per 100,000 workers. In 2014, the incidence rate for workplace fatalities across all measured industries was 3.3.[35]

In January 2016, the CDC reported on a previously unknown work-related risk to oil and gas extraction workers. These workers often manually gauge the level of fluid or collect a sample from storage tanks. Climbing to the top of storage tanks, most often alone, these workers open a small "thief" hatch to measure or collect the stored gas. This work is done outdoors and in non-confined spaces. Yet researchers found these workers are at risk for sudden cardiac death from exposure to high concentrations of hydrocarbon gases and vapors when opening the hatch. Between 2010 and 2014, nine oil and gas extraction workers suddenly died from a cardiac event while gauging or sampling. Almost all were found collapsed over the open hatch. Only one worker was wearing a respirator, which was not suitable to protect the worker from the gases and vapors.[36]

While the oil and gas boom has been hazardous for extraction workers, it has been good for business and the economy. Since 2000, North Dakota has been the epicenter of the boom, reminiscent of the gold rush of 1849. It is estimated that more than seven billion barrels of oil is sitting at the bottom of the Bakken oil fields, part of a geological formation that covers 170,000 square miles and includes a section of southern Saskatchewan and a sliver of South Dakota but is mostly centered under North Dakota.[37] With the development of fracking and horizontal drilling technologies, production in the Bakken oil fields took off in 2000, and North Dakota is now second only to Texas in oil reserves. Between 2006 and 2014, there were seventy-one deaths in the Bakken oil fields.[38] In 2012, the fatality incidence rate in North Dakota was 17.7 percent, the highest in the nation, at five times the overall national rate. In 2014, the overall fatal work injury rate for North Dakota was 9.8 percent, higher than the national rate of 3.4 percent.[39]

Although production has fallen off since, largely because of a worldwide glut of oil and a decline in oil prices, Bakken in 2015 was still spouting north of one million barrels a day. More ominously for workers, as prices decline, worker safety is further compromised as a cost-cutting measure in an attempt to pump out more profits in a declining market. Oil and gas employers routinely offer speed bonuses to workers, to produce more oil in less time. The *Wall Street Journal* reported in March 2015 that eight workers died in the Bakken oil fields since October 2014, more than in the previous twelve months, making it more appropriately known as the Bakken killing fields.[40]

OSHA has an office in Bismarck, the state capital, and is responsible for all private-sector work sites in North and South Dakota. It is a small office, with a handful of investigators. In the oil and gas industry under its jurisdiction there are reportedly more than twelve thousand active wellheads, each one a job site.[41] And, in the Wild West, anything-goes culture surrounding the Bakken oil fields, there are no OSHA regulations specific to the oil and gas industry to protect workers and rein in the industry. Instead, OSHA must rely on the General Duty Clause when it attempts to police the industry and protect workers.

Bakken is not the only place where oil and gas workers die. The *Wall Street Journal* reported in the same story that nationally, worker deaths in oil and gas are on the rise over the same time period. According to the *WSJ*, thirty-eight oil and gas workers were killed in the five months between October 2014 and February 2015, compared to sixty-eight for the entire previous year.[42]

When scientists and pundits talk about the consequences of America's addiction to oil, they are almost always talking about the environmental disaster that is occurring, resulting in melting ice caps and rising oceans. There is no dispute here about the direness of climate change. But any discussion about America's addiction to oil must include the dire health and safety consequences occurring to American oil and gas workers.

Raymundo C. Gonzalez Jr.

The name just gushes oil. Texas City. What else could it be? Located on the southwest shoreline of Galveston Bay, Texas City is a busy deepwater port on the Lone Star State's Gulf Coast. It is a major petroleum refinery and petrochemical manufacturing center. Huge tankers line up in the water waiting to be pumped full with refined sweet crude. Black gold. Texas tea. At night, a forest of oil refinery towers are trimmed with gold lights and topped off by fiery jets of burning gas, creating a hellacious vision. Texas City is home to Marathon, British Petroleum, and Valero refineries. Everyone in Texas City is no more than two degrees of separation from the oil industry. Roustabouts, pipe fitters, mechanics, truck drivers, waitresses, and health care professionals. The latter is related by necessity because petroleum refining and manufacturing is a dangerous business. Dangerous not only to the workers, but to the entire community, Texas City too often explodes, taking with it the refinery towers and workers, as well as entire city blocks and neighborhoods.

On the morning of April 16, 1947, the French-owned vessel SS *Grandcamp*, loaded heavy with ammonium nitrate produced during wartime for explosives and later used for fertilizer, caught fire and exploded. It set off a chain reaction of explosions that lasted into the next day. The entire dock area was destroyed by the first blast, including the Monsanto Chemical Company, nearby grain warehouses, and dozens of chemical and oil storage tanks. More than a thousand residences and public buildings were ruined or gone altogether. The initial explosion killed twenty-six Texas City firemen and destroyed its entire fire department and equipment. Texas City was laid defenseless.

In the end, the official death toll was 576 persons killed, with sixty-three souls burned beyond recognition. One hundred and seventy-eight persons were listed as missing, and thousands more injured. The litigation involving more than three thousand lawsuits was resolved ten years later with Congress creating a special settlement fund. As a final exclamation point to the destruction, the United States Supreme Court in 1962 refused to review a lower court decision that France, the owner of the ship, was not liable for the disaster.

Ray Gonzalez moved to Texas City when he was two years old. His father was a laborer, and there was lots of work in Texas City. There were five brothers and two sisters. They were a close family and still live in the area. While he was at Texas City High School, Ray went to work at Big Chief Food, a grocery store. He started as a sacker. It was there that he met Mary Acosta. Her mother pointed out the handsome young boy with black hair. He was shy, and it was almost a year before he got up the nerve to talk to Mary. They married in 1969, and he stayed on at Big Chief working as a meat cutter until 1971, when he hired on at Standard Oil, one of the predecessors to BP. The Texas City refinery was the third-largest in the country. Ray wanted to do better for his growing family, which eventually numbered four daughters, and Standard offered good pay and benefits.

Ray started out in operations and there learned other jobs around the plant. Eventually he became a pipe fitter. He went through all the certificate training he could. "He loved to learn," Mary said. He learned to be safety conscious from the older pipe fitters at the plant. If you got to be an old pipe fitter at a refinery you had to be careful, and lucky. Ray was an active member in the union, PACE Local 4–1, for thirty-three years. His coworkers were part of his family.

Ray was honest with Mary about the risks and dangers of his job. He wanted her to be prepared if the time came when he got hurt, or worse. So he told her stories of his coworkers' injuries and illnesses. They had a lot of conversations over the years that began with, "If anything happens to me." He wanted Mary to be prepared. A week before his incident, a friend got hurt at work. Mary and Ray had one of their "If anything happens to me" talks. He pulled Mary aside in their kitchen and gave her a list of what to do and not do. "Don't sign anything," he reminded her. He told her again that the University of Texas Blocker Burn Unit in Galveston was state of the art. It was named after Dr. Truman Blocker, who became a legend for his treatment of the burn victims from the 1947 explosion. Ray would say that a refinery was the best place to get hurt because they were expert at accidents.

The huge Texas City refinery had been around since 1934, and the equipment was aged. On September 1, 2004, Ray and two coworkers, Robert Kemp and Leonard "Maurice" Moore Jr., were assigned to do maintenance on a boiler feed-water pump. The work would contend with rust, scale, and hard-water deposits. All three men were experienced pipe fitters, with seventy-five years of combined company service. Maurice's dad, Leonard Moore Sr., and Ray had started working at the refinery together. Leonard Sr. had recently retired.

Servicing equipment with dangerous energy sources, in this case a boiler powered by hot water and steam, requires that the energy source first be isolated from the work area. This is known as "lock out / tag out," or LOTO. It is called this because the electric switch, valve handle, or other device that activates the power source must be locked closed with a locking device, or tagged out, before it is safe to work on the equipment. On September 1, the LOTO was completed on the boiler. It was believed that there was no longer any stored energy—steam and boiling water—in the pipes between the locked-out and closed valves, that the water had been drained out from the pipes that were to be repaired. The BP post-incident investigation report revealed that there was no bleeder or bypass between the valves that would have confirmed a safe energy state.

The next day, the three pipe fitters began to remove the stud bolts from the piping between the valves. The piping was around twenty feet in the air perpendicular to the floor, and scaffolding was erected to stand on to perform the maintenance. They used a pneumatic wrench, but the bolts were corroded and the threads were seized to the flange and could not be removed. They next tried using a more powerful wrench, but this too proved unsuccessful. So the bolts had to be cut off with a torch. Just before lunch, all the bolts had been cut off except three new bolts on the top flange that held up the piping. After lunch, they returned to finish removing the piping. Ray and Maurice worked on the scaffolding, and Robert was on the floor operating a crane to lower the piping. Maurice removed the new bolts, and Ray drove a wedge in between the top flanges in order to separate them. When Ray drove the wedge, the flange seals instantly broke. The pipe was not dry as expected; instead boiling water and steam were released under the force of 630 pounds of pressure. Ray and Maurice were standing on the scaffold under a spigot of boiling water and steam that was superheated to five hundred degrees Fahrenheit. They were scalded over more than 80 percent of their bodies. Robert Kemp, standing on the floor below, received burns to 60 percent of his body. Maurice was airlifted to the Blocker Burn Center in Galveston. Ray and Robert were taken by ground ambulance.

BP left a message on Mary's voice mail at home to simply call the plant. Ann, the oldest daughter, owned a hair salon in Texas City; she got a call from a client who reported that Ray was injured at work and was in the hospital. Texas City is a refinery town where everyone knows everyone. The kids go to school and grow up together. Lives are interconnected at the refinery like so many pipes, and bad news travels fast. Ann called Mary. The drills Ray practiced and imbedded in Mary and the girls took over. Mary said, "I knew not to panic. The main thing was to get to Ray at the hospital. All the girls took over their duties. No one panicked."

Maurice died the next day. Robert eventually recovered at home from his injuries. For Ray, he endured for two and a half months at the burn center. When Mary arrived at the hospital, he was alert, and she was able to talk to him before he was intubated to help him breathe. He did not want to be on life support.

Skin is the largest organ in the body. It protects everything underneath. Ray had third degree burns. His layer of skin was seared off. He had no skin to graft. So, initially, cadaver skin was used to cover his body while Ray's own skin healed and grew back. Later, the doctors shaved his head and used patches of scalp skin for graft. In all, Ray had more than ten skin-graft operations. Each surgery was risky and knocked him backward, Mary said. During one surgery Ray had a heart attack. The grafts were open wounds, and there was enormous risk of infection. Ray was on a breathing tube and couldn't talk. When Mary and the girls visited, they had to dress in Hazmat suits. It made the visits less intimate. Despite it all, the infection set in and took hold. Ray's organs began to shut down. In the end, Ray was on a ventilator to keep him alive. Mary said, "It was wearing him down." He was suffering. Mary took him off the ventilator, and he died on November 12, 2004, of organ failure.

Throughout the ordeal there was a lot of support, including from BP. Ray's union family "was there for us." They helped explain in plain terms the cause of the accident. They reviewed the results of the internal investigation with Mary and her daughters. There were many causes. All of them were avoidable. There was an obstruction that prevented one of the valves from fully closing. The LOTO procedures were inadequate. The design should have included a bypass around the valves. OSHA found six serious violations, which is where there is "a substantial probability that death or serious physical harm could result." In addition, one violation was deemed willful. OSHA defines willful as "a violation that the employer intentionally and knowingly commits." The definition goes on to state that "the employer is aware that a hazardous condition exists, knows that the condition violates a standard or obligation of the Act, and makes no reasonable effort to eliminate it." Willful violations may result in criminal proceedings where death has resulted, with fines and imprisonment of "up to 6 months." Two convictions, for two deaths, double the term of imprisonment. No criminal indictments were sought against BP. Instead, BP was fined $102,500 for the "incident" that took the lives of Ray Gonzalez and Maurice Leonard Jr.

The following spring, on March 23, 2005, there was a refinery explosion at the BP plant in Texas City. This time, 15 workers were killed and 170 injured. BP was in the process of contesting all the penalties for Ray's and Maurice's deaths when the refinery blast occurred. Apparently, BP thought better of claiming that it wasn't at fault as the death toll in Texas City added fifteen more men. Since 1971, when records were kept, there were at least nine other accidents at the refinery that resulted in serious injury or death. In late 2012, BP sold the Texas City refinery to Marathon in a deal worth $2.5 billion. It was part of BP's strategy for growth and to sell assets to help fund the millions it paid for the Deepwater Horizon oil spill and to the victims and families of the 2005 explosion.

For Mary Gonzalez and her family, there was little comfort or satisfaction in BP's fortunes. She receives monthly workers' compensation benefits, which are a percentage of Ray's wage. And just to be sure she isn't receiving any money she isn't entitled to, state investigators regularly check in on Mary to see if she has remarried or is cohabiting. Mary said this invasion "just adds another layer of pain" to her open wounds.

Katherine Rodriguez, Mary and Ray's second daughter, described the family after Ray's death as "numb and dead inside." But although the grief never goes away, Katherine and her sisters knew what their dad would have wanted. "He wouldn't want us to be sad or to grieve for long periods. Dad would want us to do something to prevent this from happening again. We were his voice," Katherine said. "I promised this to him in the hospital."

Katherine became involved with a survivors' network, the USMWF. They pushed OSHA for more outreach to family members. With a change in administration in 2009 came a change at OSHA. "They listen now," said Katherine. "They listen to us, to our stories." Ray's picture hangs in OSHA's conference room in Washington, DC, along with those of others who lost their lives at work.

Dredging

> Sometimes we'd have that whole river all to ourselves for the longest time. Yonder was the banks and the islands, across the water; and maybe a spark—which was a candle in a cabin window—and sometimes on the water you could see a spark or two—on a raft or a scow, you know; and maybe you could hear a fiddle or a song coming over from one of them crafts. It's lovely to live on a raft.
>
> MARK TWAIN, ADVENTURES OF HUCKLEBERRY FINN

The most important American waterway is the Mississippi River, which courses through the heart of the continent, flowing through or past ten states on the way from its headwaters in Lake Itasca, Minnesota, to the

Gulf of Mexico. Its watershed is surpassed in size only by those of the Nile, Amazon, and Yangtze Rivers. Tributaries from thirty-one states and two Canadian provinces flow into the great river between the Rocky and Allegheny Mountains. There are twenty-nine locks and dams on the Upper Mississippi, a nine-foot deep shipping channel from Louisiana to Minnesota, and great ports along or near the Mississippi River that serve as the off ramps for America's commerce.[43]

Altogether, America has more than twenty-five thousand miles of inland and intracoastal waterways that carry waterborne traffic loaded down with millions of tons of coal, petroleum, chemicals, iron ore, lumber, grain, sand and gravel, manufactured goods, and other commodities. There are hundreds of aging locks and dams that allow barges to float along the uneven rivers. The thousands of miles of rivers, channels, and canals connect 926 coastal and Great Lakes harbors and ports. What's at stake is the safe passage of more than 2.2 billion tons of commerce that is shipped along America's waterways and into its ports by thousands of barges, tankers, and container ships.[44]

In addition to the commercial waterways, our beaches and coastlines, internal and at the continental edges, must be maintained and replenished for recreational purposes and to act as a protective barrier for homes and businesses from erosion and the devastating incursion of seawaters in times of storms. Finally, wildlife and fragile ecosystems that live along our beaches and wetlands must be protected and restored.

The tasks of keeping America's massive waterway systems open and navigable, our homes and businesses protected, and our wetlands restored falls to the U.S. Army Corps of Engineers (USACE) and its private contractor partners. This most often involves dredging, to create or enlarge shipping lanes and ports and to maintain existing waterways from sediment buildup. As boats become bigger, shipping lanes and ports have to be deepened and enlarged to handle their draft and size. At the end of 2013, Shell Oil launched the *Prelude*, a Korean-built ship that, if stood on end, would be taller than the Empire State Building. To call it a boat is misleading. Shell refers to the *Prelude* as a facility, or an FLNG, a floating liquid natural gas facility. The *Prelude* displaces more than six hundred thousand tons of water, five times the displacement of the largest aircraft carrier.[45] These mega-ships require deeper and larger shipping lanes.

For fiscal years 2008 through 2012, USACE reported that an average 212 million cubic yards of material was dredged from America's waterways each year.[46] Most of the dredging was performed by private contractors hired by the USACE at a cost to taxpayers of more than a billion dollars. How safe is this work?

Dredgers are specialized machines, typically installed on vessels, or sometimes working from land, that excavate the bottoms of rivers, harbors, and shallow coastal waters to keep them navigable and clear from silt, or to deepen already existing channels and harbors. The two main types of dredgers are mechanical and hydraulic. A mechanical dredger is not unlike a land-based excavator, and scoops out the hard-packed material with a backhoe, bucket, or grabber. A hydraulic dredger, on the other hand, uses a pump to dislodge or loosen the material and suction it in a liquid form, known as slurry. The dredged material is then deposited into floating hoppers or barges for disposal.

Most dredgers are custom built and specially fitted with the machinery to safely excavate and remove the sediment and other material. However, some dredging operators may cut corners and rig a standard backhoe or crane onto a pontoon or barge. All workers on dredgers are at risk for a bucket to malfunction, dumping the excavated material onto a worker. Drownings are common for dredge workers. A worker may simply slip and fall off the dredger or barge, only to be found downriver hours later. Dredgers may also capsize, and workers can drown in those events. All workers on a dredger should wear a lifejacket or other personal flotation device (PFD). But, owing to a lack of training, or a lack of PFDs, many dredge workers are unnecessarily at risk for drowning. The USACE recently launched a safety campaign titled "Life Jackets Worn . . . Nobody Mourns."[47] In addition to drownings, dredge workers suffer from other common injuries, including from musculoskeletal disorders, amputations, crush injuries, fires, and explosions.

Coast Guard data reported that between 1997 and 2006, 305 employees were killed on barge/tow combinations, and 379 explosions or fires occurred on barges or towboats, killing fourteen employees.[48] Notwithstanding this record, for dredge and barge workers, OSHA could propose only "guidance" that addresses falling overboard, the major cause of injury and death. But "guidance" is not like a final regulation that can result in an enforcement action against a violator. The OSHA guidance manual makes this unmistakably clear in its disclaimer: "This guidance is not a standard or regulation, and it creates no new legal obligations. It is advisory in nature, informational in content, and is intended to assist employers in providing a safe and healthful workplace."[49]

OSHA has been unable to promulgate any new regulations that might help protect dredge workers. Final regulations for slips, trips, and falls, as well as for confined spaces and crane operators, have been in regulatory limbo for years. Each rule could protect dredge workers from serious injury and death, but they remain hostage to political gridlock. And all the while, dredge and barge workers continue to get injured and die in preventable incidents.

Scott Shaw

Philadelphia is washed on three sides by the Delaware and Schuylkill Rivers, which made the city a great commercial seaport. The Schuylkill River (pronounced "Schoolkill") flows for around 135 miles. It was named by Dutch sailors and means "hidden river." It is said that it was given this name because its mouth could not be seen by mariners sailing up the Delaware until the two rivers actually joined. It begins its journey by combining two branches: the East Branch descends from the Appalachian Mountains and joins the West Branch at Port Clinton, Pennsylvania. On its way down to Philadelphia, the river glides past Reading and Valley Forge. Together with the now-defunct Reading Railroad, the river carried coal down from the mountains to the towns along the Schuylkill, before reaching its mouth at the Delaware River at the site of the Philadelphia Naval Shipyard. Since 1812, the Schuylkill has been providing fresh drinking water to Philadelphians, courtesy of government. The city's Fairmount Water Works was the first of its kind in the United States. In the 1800s, wooden ships, and later steamers, from all over the world sailed into the City of Brotherly Love, bringing wares and treasure, while loading Pennsylvania coal into their holds. The city's naval yards have been building, refitting, and repairing ships since 1776.

In order to keep the shipping lanes open and clear, the Schuylkill has to be constantly dredged. Rivers naturally carry suspended soil and sand as they flow toward the ocean, depositing sediment and creating sandbars and shoals along their paths. In turn, the river bottom becomes shallower than shown on nautical charts, grounding boats. Cargo and the boats may be damaged, and in the worst cases, environmental disasters can occur if the cargo spills into the waterway. Larger ships today, stacked nine tiers high on the deck with shipping containers, need deeper shipping lanes to accommodate their draft. Dredging is like alluvial angioplasty. Without it, the heart of Philadelphia stops beating.

Scott Shaw was thirty-eight years old when he was killed. He was tall and athletic. He had a handsome, broad face, framed by blond curls and set off by blue eyes. In high school he was on the baseball team, and in the summer he was a lifeguard. Scott was a strong swimmer. In 1981, he graduated from high school across the river in New Jersey. He got married to his high school sweetheart, Kelly. He started to work in construction. Scott loved working with heavy machinery. Bulldozers and excavators. Soon they had a son, Scotty. The marriage didn't last long, and Scott and Kelly got divorced.

Construction was seasonal, and in the winter Scott was often laid off. He passed the time cooking and following his beloved Philadelphia teams—the Eagles, the Flyers, the 76ers. In the summer, it was the Phils. He met Holly, and they got married in August 1993. Holly has been a schoolteacher for twenty-three years at the E. M. Stanton Elementary School. Stanton was Abraham Lincoln's secretary of war. Holly now runs the school's computer lab.

Holly and Scott had two boys together, Ryan and Nicholas. Ryan was five when Scott died. Nicholas celebrated his third birthday five days after Scott's death. Neither remembers much about their dad. Holly has tried to create memories for the boys. Ryan mostly remembers fishing for catfish at a spot on the Schuylkill, behind a Walmart. They had to crawl under the fence and used bread and hot dogs for bait. On Father's Day and Scott's birthday, Holly and the boys would release balloons over the river from the fishing spot. She stopped bringing the boys there because it was too sad for them.

In 2001, Scott changed jobs. He left construction and started to work for the Armco Construction Company, which had contracts with the U.S. Army Corp of Engineers to dredge the Schuylkill. Armco leased barges from Weeks Marine. Dredging crews generally use two barges. One barge dredges and pumps the material into the other barge. The barges are supposed to be connected by a sturdy platform or gangway so that the workers can safely move from one barge to the other. The barges leased from Weeks were connected by old tires, strung together in an unsteady web between the barges. Traversing the barges was made all the more difficult by the fact that the deck heights were different.

A few years after Scott died, barge workers from Staten Island went on strike because Weeks's barges there were connected by old tires. A worker was crushed to death between the two barges. Weeks denied any fault or responsibility for the tire platform.

Holly said that Scott had come home more than once soaking wet after falling into the river from the Weeks barges. He didn't complain or go on strike. He was a good swimmer, and he dried off and went back to work.

On September 7, 2002, Scott and two other workers were cleaning the barges after finishing a dredging project near the Girard Point Bridge. According to the OSHA Accident Investigation Summary, two workers were pumping dredged material from a hopper barge onto an Army Corps disposal site on the shore of the Schuylkill. Scott crossed over the tires to the other barge. When one of the workers looked up a moment later, he was gone. All that was left was his baseball cap floating in the water. After doing a quick search of the barges and the water around them, the workers called for help to a nearby company. Scott's brother-in-law, Craig, worked there and answered the call for help to look for a missing barge worker. Craig's company used its skiff to search the river around and between the barges. Around an hour after Scott went missing, the Philadelphia Police Marine Unit was finally called. They arrived and used grappling hooks to search for Scott.

September 7 was a Saturday, and Holly was home with the boys. At around 11:30, Holly called Scott's cell phone. She doesn't remember why. On the barge, the workers customarily placed their cell phones and wallets in a plastic bag, just in case they fell into the river. When Holly called, a coworker answered and said, "Scott's not here now." He didn't tell Holly that Scott had been missing for over an hour. Holly immediately knew that something was wrong, but waited for Scott to call her back. By the early afternoon, Scott's coworkers and Armco's owner came to the house. Holly remembers opening the door and seeing them standing there. She sank to her knees when they told her that Scott was missing. Before they left, one of the workers told Holly not to call any lawyers. After they left, Ryan came downstairs. He also knew immediately from seeing Holly that something was wrong. She told him, "Daddy fell in the river and they're looking for him." Holly called her family, who soon all congregated at the house in a vigil, waiting for word from the police. They were supposed to search for only forty-eight hours, but said that they would keep looking until they found Scott. Ryan started drawing pictures of Scott in the water drowning. On Monday night, the police found Scott's body under the Girard Point Bridge. They told Holly that he sank to the bottom of the river after falling from the barge, but eventually rose to the surface from gases released by his decaying body.

After September 11, 2001, Scott and Holly had talked about what to do if one of them died. Scott said he wanted to be cremated. His ashes, and his father's, were spread in the Pinelands in New Jersey, a reserve area with biking and hiking trails. There was a service for Scott at a funeral home in South Philadelphia. Work boots with flowers lined the steps to the home.

Holly was in shock from losing Scott. She sent Ryan back to school on the one-year anniversary of 9/11. There was a lot of death around her. Ryan told his first-grade class about Scott dying when they talked about 9/11. Holly took a leave from teaching for a couple of years while she struggled to put her life back together as a widow and single mother. She told the boys, "We have to laugh every day," while she told herself to get through ten and twenty minutes at a time. Ryan went to grief counseling. Holly went to counseling herself. She reconnected there with a high school friend whose husband was killed in a car accident a week before Scott died. Holly says that "time doesn't heal. It just helps you to cope."

Holly received monthly survivor death benefits from the state's workers' compensation fund. The benefits were on the condition that she didn't remarry. She could get the survivor benefits if she suffered alone as a widow raising two young boys. For a while after Scott died, investigators from the state would periodically check in on Holly to see if she had remarried. Holly hired an attorney to help her navigate the workers' compensation system. She eventually agreed to a lesser lump-sum settlement. That way she could remarry if she wanted to. The money is in the bank for the boys' college. Holly also tried to sue Armco, but they declared bankruptcy. Instead, she sued Weeks Marine for negligence. Weeks denied any responsibility or knowledge about the tires. She eventually settled with Weeks for $5,000.

OSHA opened its investigation on September 9 and closed it six days later. OSHA issued six "serious violations" against Armco, including one for not having procedures to make certain that barge workers wore life vests. Armco was also cited for using tires as the means for moving from one barge to the other. All totaled, Armco was fined $4,950. Armco protested the fines, and they were abated to $4,000. OSHA never told Holly about the fines or the abatement. She was never given the opportunity to tell her side of the story. She was never allowed to tell OSHA how their lives had been torn apart by Scott's needless death. A murder victim's family is allowed to speak at a parole hearing. But a family who are survivors of a workplace death don't have any rights. They don't exist for purposes of investigating and judging their loved one's death. They have no rights.

Holly got remarried to Bill Hollis. Bill has two grown sons from a previous marriage and is also an avid sports fan, although of the Dallas Cowboys. No matter. He makes Holly laugh every day. More important, Bill has been very supportive of Holly's efforts to keep Scott's memory alive, especially for the boys.

In trying to learn about Scott's death and the OSHA investigation, Holly became involved in PhilaPOSH, the local chapter of a national worker safety organization. She is now on its executive board. Every April, on the last Friday, there is a national and international Workers' Memorial Day. PhilaPOSH, local labor unions, and other community groups gather for breakfast and then march together wearing placards bearing the name of a worker who was killed in the prior year. At the end of the march there are speakers, and the names of workers killed that year, and how they were killed, are read aloud. Holly also joined United Support and Memorial for Workplace Fatalities, a national support group for survivors of workplace deaths. Through the USMWF, Holly has testified before the U.S. Senate Committee on Health, Education, Labor and Pensions, urging passage of laws imposing greater penalties on employers, especially on repeat offenders. She has advocated on behalf of the Protecting America's Workers Act, which would require OSHA to involve families early on in the investigation and penalty phase. Holly says that it is important for family members to have a say on making sure that there is accountability for the death of a loved one. It is only fair. Not surprisingly, Congress has not acted on this proposal.

Logging

> In those days, much of North America was filled with thick, green
> forests. Paul Bunyan could clear large wooded areas with a single
> stroke of his large, sharp axe.
>
> THE STORY OF PAUL BUNYAN (ADAPTED BY GEORGE GROW)

As a small child, I loved the story of America's iconic lumberjack, Paul
Bunyan, and Babe the Blue Ox. I read the story so often, I had it practically memorized. And I was mesmerized by the part where Paul Bunyan
dug out the Great Lakes to provide drinking water for Babe. Of course,
I could practically taste the hotcakes, each one so large it could feed five
giant lumberjacks. This tall tale is part of the folklore of America. But for
real flesh-and-blood lumberjacks, the work is anything but fantastical.

Ever since the first settlers landed in Jamestown, logging has been in-
extricably twined with the growth of America. From forests rich in pine,
cedar, chestnut, and black and white oaks, logging built America's cities
and towns, and its boats. It provided needed structural support for underground coal mines, and the railcars and rail ties on the lengthening lines
crisscrossing the continent.

Logging is also America's most dangerous job, well ahead of other dangerous occupations. The job classifications of loggers are colorful and,
unlike those of the high-tech and Internet-based industries of today, evoke
America's frontier past with names like "fallers," "buckers," "tree climb-
ers," "rigging slingers and chasers," and "log sorters and graders." But
these jobs involve dangers such as scaling tall trees with power saws and
other heavy tools, while harnessed by a rope. The fallers go at trees with
powerful chain saws, and hope that they have correctly judged the lean
of a tree and the wind before making the final backcut that fells the tree.
Explosives are often used to clear obstacles, so the handling and storage
of these hazards are critical components of any safety plan. The stacking,
sorting, and shipping of massive tree trunks headed for the sawmill or
pulp mill pose significant dangers, from being rolled over and crushed by
a runaway log, to truck accidents caused by an unstable load while headed
down a narrow mountain logging road. And, finally, the sawmill has its
own dangers, particularly while pushing a log back and forth through a
large circular or band saw as it cuts the raw timber into planks, as well as

dangers from inhaling sawdust and from chunks of flying wood jettisoned while debarking a tree.

OSHA's "Logging eTool" website describes logging as "the most dangerous occupation in the United States," and not just because of the hazardous tools and equipment used, such as chain saws and logging machines. In addition, loggers "deal with massive weights and irresistible momentum of falling, rolling, and sliding trees and logs," and other hazards and conditions beyond their control, "such as uneven, unstable or rough terrain; inclement weather including rain, snow, lightning, winds, and extreme cold or remote and isolated work sites where health care facilities are not immediately accessible."[50]

According to the CDC, in 2010 the logging industry employed 95,000 workers. In that year, there were 70 industry deaths, resulting in a fatality rate of 73.7 per 100,000 workers, or twenty-one times higher than the average fatality rate for all American workers in 2010.[51] This rate has only increased, and for 2014, BLS reported that the fatality rate for loggers had climbed to 109.5, compared to a rate of 3.4 for the average worker.[52]

The Forest Resources Association, the industry trade group, maintains a website that lists and reports on serious injuries and fatalities to loggers.[53] Falling trees and limbs appear to be the most likely cause of injuries and fatalities. But being crushed by logs, or by heavy equipment and trucks, is also a common occurrence. So is electrocution, and power saw malfunctions that send the blade or chain into an unsuspecting logger. And (as OSHA noted) loggers perform these hazardous jobs at remote locations, often far from needed medical care—and the median pay in 2015 was $17.41 an hour.[54] The increasing number of fatalities, giving logging the dubious distinction as the most dangerous job, has been attributed to inexperienced workers and poor safety training. Worker inexperience is a common theme in America's changing workforce. But it has especially deadly consequences in the logging industry.

America's romance with lumberjacks still endures. There are flapjack eating contests, and amusement parks with rides that glorify loggers. Then there are the Lumberjack World Championships, televised on ESPN, in which muscular men and women compete in pole climbing, log cutting, log rolling, log tossing, and other contests. But the reality of this most dangerous work is, like that of many other jobs in America, hidden or ignored in favor of fantasy and lore.

Sherman Holmes

People do not often know how they will die. When it is unexpected and out of sync with the life they lived, such a death is difficult for the survivors, who are left to puzzle how loved ones died the way they did. When Nicole Boone received a telephone call from her cousin living up in northern Michigan, informing her that her father Sherman Holmes had died in a logging accident, she thought it was a joke.

Tustin, Michigan, is located in part of the great swath of white pine forests stretching from Maine through New York and over to northern Michigan, which since 1840 has fed America's seemingly inexhaustible demand for lumber. North of an imaginary line from Muskegon to Saginaw, the rich pine forests grew. White, jack, and red pines. But white pine was king of the lumber industry in this region of northern Michigan. That's where Tustin and other towns like it are dotted among the forests and lakes and rivers. At the last census, there were 230 persons living in Tustin. Most of the original settlers were German, English, or Swedish, people who came to homestead and work the forests. Sherman Holmes's grandfather came from Germany. He was a logger. He had thirteen kids who in turn had bunches of children. Sherman was the third of five children and grew up on a small farm. There were lots of cousins in the area and lots of people named Holmes. Tustin has a sawmill owned by Alan Holmes, a distant relative. Sherman's father, Doyle, was also a logger, as was Sherman's brother Ves. Sherman learned how to log from his grandfather and father. Since he was old enough to walk, Sherman Holmes spent most of his time in the woods hunting, fishing, and logging. He could look at a tree and tell you which way it would fall when cut.

In 1973 Sherman graduated from Pine River High School in nearby Leroy, and he returned there in 1978 to work first as a bus driver and then for most of his career as Pine River's maintenance man. He retired in May 2010 after thirty-five years. He worked at the school for health insurance and a pension. All the while, he ran Sherman Holmes Forest Products, essentially a one-man lumbering business. He sponsored Nicole's softball team.

His business model was simple. He would bid to cut trees with a landowner, then sell what he could cut to the local sawmill, and split the proceeds with landowner as they had agreed. He owned an iron mule, which carved a path in the forest to where the trees were. The mule could hoist the felled and limbed trees onto the back of Sherman's small flatbed logging truck. He was once hired by John Glenn, the astronaut and senator, to thin out the white pines on property he owned in northern Michigan. Sherman worked at the school on weekdays, and on weekends and during the long summer days he also worked as a lumberjack. Nicole remembered her father always working. If he wasn't working, he was hunting or fishing. Sherman was also a Tustin area volunteer fireman. He did not sit still. She also remembered his last car: a 2001 Pontiac Grand Am. It smelled of sawdust. Logging was in his blood. But in the end, it became too expensive to run his business. The costs of maintaining the flatbed and iron mule were too much, and he shut down Sherman Holmes Forest Products in 2002 but continued to work odd jobs logging for friends and others.

After retiring from the high school in July 2010, Sherman was hired on full time at K and K Forest Products LLC, a small logging and trucking business in Leroy. It is owned by two brothers, Ken and Kevin Proctor. Like the Holmeses, the Proctors have lived in northern Michigan forever, and the families' lives have often crisscrossed. Ken and Sherman went to school together. Around thirty years ago, Clifford Proctor, a relative, drowned as a young child in a farm pond. The family didn't have enough money for a plot in a cemetery. The Holmes family had eight plots in the Burdell Township cemetery. Sherman gave away one of his plots to Clifford Proctor's family. Sherman is now buried next to Clifford.

Sherman did not like working for K and K. When he came to visit Nicole in Ohio for Thanksgiving in 2010, he had a knot on his head, and his face was bruised. He complained that the crew he was working with did not work very hard and was not very experienced. A tree fell and caught Sherman in the head. He was lucky it wasn't worse.

February 1, 2011, was a clear blue but windy day in and around Tustin. The average reported temperature was sixteen degrees. It was the kind of deep cold that stiffened your clothing and froze your snot. A snowstorm that would eventually dump eighteen inches on top of the already deep snow was expected that evening. In other words, it was a picture postcard winter day in northern Michigan.

Sherman was not scheduled to work and used the time that morning to get his taxes prepared. Nicole remembered that was odd, as he was a dedicated procrastinator, especially when it came to his taxes. But on that day, for whatever reason, Sherman wanted to get his financial affairs in order early. While running errands that morning he got a call to come in to work. K and K was short employees. The crew was working in Evart, about thirty miles south and east of Tustin. They were hand felling trees on private land, clumped aspens and other hardwoods. Clumped trees share a common stump that cannot be felled by mechanical means or processor. They are cut by handsaws.

According to the Osceola county sheriff's incident report, while a co-worker cut a notch in a tree (which determines the direction of the fall), Sherman was standing behind him. The man may not have seen Sherman, but he wasn't hard to see. Sherman was wearing a bright orange bib over-all snowsuit, but no safety helmet. Nonetheless, when the coworker made the back cut and the tree began to fall, he noticed Sherman running in the deep snow, away from the falling tree. At the last second, Sherman threw away his chain saw and slipped or dove into the snow just before the tree fell on top of him. The tree looked like a giant three-pronged fork. It was approximately seventy-five feet high and twenty inches in diameter. Sherman was struck in the neck and upper back by the Y of the tree, around fifty feet from the stump. His coworkers had to cut the tree off from atop Sherman. They rolled him over in the snow and onto his back and called 911. He was still breathing but unconscious, with blood oozing from his mouth and left ear. After a few more minutes, he stopped breathing. Sherman Holmes was fifty-five when he died.

The photographs taken by the sheriff of Sherman lying in the snow almost belied the brutality of his death. He looked peaceful in the white snow, like a young boy making a snow angel. You have to look closely to see the trauma to his head. The forensic pathologist reported a "craniocerebral trauma" and "diffuse subarachnoid hemorrhage" caused by a "linear fracture" of the cranial fossa. The falling tree crushed his skull. The Michigan Occupational Safety and Health Administration arrived on the scene that afternoon and opened its investigation into Sherman's death, but left after a short while because of the impending snowstorm. MIOSHA returned two days later and made a physical examination of the accident site, and took some photos of the equipment and the fallen killer tree, now all covered by new snowfall. The MIOSHA investigator recommended citations for violations of the Michigan Occupational Safety and

Health Act. Specifically, MIOSHA issued citations for not providing to each employee on the work site at no cost a protective helmet, and ensuring that employees did not cut down trees without wearing a helmet. The proposed penalty for this "serious" type of violation was $125. The cost of a protective helmet with a face screen and hearing protectors is $44.95 on Amazon. A deluxe model is under $100. Sherman Holmes's life is, as the commercial touts, priceless.

The second and third "serious" violations were for not providing minimum training and because Sherman was within fifty feet of his coworker who was felling trees. This regulation requires that employees shall be spaced "not less than 2 tree lengths of the trees being felled." According to the regulations, before beginning a cut, "a feller" shall check for the location of all other employees, wind conditions and "lean of the tree," and plan and "clear a path of retreat" in the event the tree doesn't fall as intended. In fact, before starting a back cut, "a feller shall determine that the impact area is clear of other employees."

The employer admitted to the MIOSHA investigator that it needed to train all its employees. Too late for Sherman. K and K abated the violations by now requiring mandatory monthly safety meetings, which include emphasizing the requirement of wearing personal protective equipment and reviewing the safety standards for felling trees. An unmentioned problem that day was the wind, which may have affected the intended direction of the falling tree. The regulations require that "work shall stop" under conditions of high wind.

The proposed fines were $700 for each of these violations. K and K paid a total fine of $1,525 less than three months after Sherman was killed. No compensation of any kind was paid to Sherman's family.

In its letter of condolence to Nicole, MIOSHA acknowledged the obvious, "that each worker's death is a deep personal loss and tragedy." It further "committed to do everything possible to prevent similar types of accidents." But is it so "committed" when the value of these three "serious" violations costing one man his life is $1,525? The mathematics of deterrence simply do not add up.

In the end, her father's unnecessary and untimely death has made Nicole, an Ohio schoolteacher, a citizen worker safety advocate. She has testified before Congress and has met with White House staffers to urge action on commonsense injury and illness prevention programs that are being delayed in Congress while thirteen workers die each day. Most important, Nicole is trying to give a voice to her father and other workers who have died, voices that for too long have been drowned out by the roar of more powerful and deep-pocketed interests.

Sherman Holmes had survived in the woods for fifty-five years, most of them spent logging and often alone. He would have been as surprised as Nicole to learn that he died in a logging accident. By a falling tree, no less. He surely knew its dangers, and was just as certain on how to avoid them. What he did not and could not know for sure was whether his coworkers knew the same. It is also uncertain whether the wind was a contributing factor, or whether a safety helmet, even the most expensive, would have prevented his skull from getting crushed. But for Sherman, that clear, cold morning was a perfect killer storm, one that repeats itself at millions of work sites across America: the willful disregard of simple, lifesaving rules and regulations. Not all workplace fatalities can be prevented. Logging is hazardous business—there is no dispute of that fact. But the odds of a serious or fatal accident increase manyfold by such disregard. Of that we can be certain.

Combustible Dust

> All the world is made of faith, and trust, and pixie dust.
>
> J. M. BARRIE, *PETER PAN*

For Tinker Bell in J. M. Barrie's *Peter Pan*, pixie dust was a magical glittering powder, which when combined with happy thoughts gave flight to Wendy, Michael, and John to Neverland. Not so for American workers. Dust in the workplace is a silent and deadly killer. The dust sprinkling almost every workplace will, under the right conditions, explode into massive fireballs, destroying property and killing anyone unlucky enough to be nearby.

The Imperial Sugar Company operated a sugar refinery in Port Wentworth, Georgia. There, Imperial refined raw sugar into granulated sugar. Imperial is one of the largest sugar refiners in the United States. At its Port Wentworth plant, Imperial processed over seven hundred thousand tons of sugar each year for industrial bakeries and large and small grocers.

At around 7:15 p.m. on February 7, 2008, a violent explosion occurred in a silo that stored granulated sugar. According to the Chemical Safety Board's (CSB) investigation report, a series of explosions and massive fireballs erupted at the plant.[55] The CSB investigators described that the "three-inch thick concrete floors in the south packing building heaved and buckled from the explosive force of the sugar dust-fueled explosions," while the roof was shattered and blown off.[56] For many of the hundred or so workers, there was no warning or opportunity to escape the inferno, as walls, equipment, and furniture were tossed around. The CSB reported that "superheated air burned exposed flesh. Workers attempting to escape struggled to find their way out of the smoke-filled, darkened work areas. Debris littered the passageways. Some exits were blocked by collapsed brick walls and other debris. The fire sprinkler system failed because the explosions ruptured the water pipes."[57] The violent fireballs continued to explode for more than fifteen minutes, fueled by spilled sugar and sugar dust. In the end, nine workers were killed from the initial explosions and fires, while six later died at the Joseph M. Still Burn Unit in Augusta, the last victim suffering for six months. Thirty-six workers were injured, some with severe burn injuries. It took the fire department a week to finally put out all the smoldering silo fires.[58]

Combustible dust explosions have been a known safety hazard as early as the late eighteenth century, when an Italian count reported an explosion in a baker's flour warehouse. In the United States, the Washburn "A" Mill in Minneapolis, then the largest flour mill in the world, exploded on May 2, 1878, into a fireball killing all fourteen workers. The CSB investigation at Imperial Sugar noted that combustible dust hazard studies, including for sugar dust, went back more than eighty years. The CSB included in its investigative report a quote from a 1925 publication, *The Dust Hazard in Industry*, by William Gibbs, which concluded that "sugar, dextrin, starch and cocoa are the most dangerous, sugar exceptionally so. Sugar ignites when projected as a cloud against a surface heated to below red heat, and when ignition has taken place, the flame travels throughout the dust-cloud with great rapidity."[59] In 2006, the CSB issued a 108-page report simply titled, *Combustible Dust Hazards Study*. The report detailed the history of 281 major combustible dust explosions from 1980 to 2005, which resulted in 119 worker deaths and 718 injuries.[60] The purpose of the study was to raise awareness of the need for a combustible dust safety standard. Yet, a decade later, there still is no OSHA combustible dust safety standard on the books.

After the Imperial explosion, legislation passed in the House in 2008 by a bipartisan majority, directing a combustible dust safety rule. Known as the Worker Protection against Combustible Dust Explosions and Fires Act of 2008, the legislation required that within ninety days after its enactment, "the Secretary of Labor shall promulgate an interim final standard regulating combustible dusts."[61] No more fooling around with workers' lives. Except the House bill never made it out of the Senate. The 2010 midterm elections occurred, and bipartisanship went up in flames. The history of combustible-dust rule making, as reported below by the CSB's "Rulemaking Watch" website in November 2014, followed a depressingly dead-end path: "In April 2009, the Occupational Safety and Health Administration (OSHA) announced that it was initiating a comprehensive rulemaking on combustible dust. Shortly thereafter, in October 2009, OSHA published an Advanced Notice of Proposed Rulemaking (ANPR) announcing the agency's intent to promulgate a dust standard, and requested comments, 'including data and other information, on issues related to the hazards of combustible dust in the workplace.' "[62]

In December 2009, February 2010, and April 2010, OSHA held stakeholder meetings in Atlanta, Chicago, and Washington, DC. In June 2010, the agency held a Virtual Stakeholder Meeting to collect additional information. The agency also opted to convene an Expert Forum in May 2011.

The next required step in the federal rule-making process is the convening of a Small Business Regulatory Enforcement Fairness Act (SBREFA) panel. OSHA has delayed the SBREFA panel since 2010. In May 2016, the CSB website reported that the SBREFA panel was postponed until October 2016.[63]

In the midst of this regulatory inaction, the CSB documented fifty combustible dust incidents from 2008 to 2012, leading to 29 deaths and 161 injuries.[64] Dust explosions are preventable with proper engineering controls, ventilation, training, and other measures. Yet continued inaction and inexcusable delay by those responsible for protecting our workers predictably result in more deaths, which could have been preventable. There are no fairy tale happy endings for these workers who work at unnecessary risk.

Shawn Boone

By all accounts, Shawn Boone was a happy guy. In a photo posted on a website memorial, Shawn is standing on his beloved farm getting ready to do some chores, wearing a Carhartt coat, work gloves, and a red stocking cap. Mostly, he wore a big grin. "He never had a bad day," said his sister, Tammy Miser. That is, until he did.

Shawn was the third of four kids for Hope Mock. Three boys, and Tammy, the oldest. There was no father around. The family lived in Disko, Indiana, a tiny unincorporated farming community due west of Fort Wayne. Tammy said that there was disagreement among those who lived there on whether the town was spelled with a *c* or a *k*, a dispute apparently still unresolved, according to the few Google search results on the subject.

Shawn was always doing and fixing something. He had a huge garage on the farm and loved to tinker with whatever was broken. Accomplishing things even as simple as quickly solving a Rubik's Cube was important to Shawn. When he died, Shawn was going to trade school to learn how to be a machine operator. He always wanted to better himself. Shawn never married and had no children of his own. He did have Duchess, however, his English bulldog. Shawn was close to his family. He taught Tammy's oldest daughter how to drive and bought her her first car.

Hayes Lemmerz International Inc. designs and manufactures aluminum and steel wheels for cars and trucks, and has annual sales over $1 billion. Tammy's husband, Mark, worked at a Hayes Lemmerz facility in nearby Huntington, which made cast aluminum alloy wheels. He recommended Shawn for an open position in maintenance. The Huntington facility is located alongside other manufacturing facilities in a light industrial park. The entire facility—including the foundry furnaces, casting machines, and finishing equipment—was housed in a single, 220,000-square-foot steel-frame, steel-clad industrial building.

In 2001, Hayes filed for Chapter 11 bankruptcy protection. While in Chapter 11, Hayes began to downsize at the Huntington plant. Tammy and Mark looked for work elsewhere, and eventually moved to Lexington, Kentucky, in 2001. Shawn stuck it out and stayed behind to work at Hayes.

The aluminum wheels are rough cast, then finished by machining, coating, and polishing. These machining activities create scrap aluminum.

Aluminum scrap is processed and melted using a chip-melting system. The flow diagram for the chip-melting process is a complicated series of steps, with machinery designed to dry the surfaces wet from the cutting-oil residue, using a rotary kiln, like a clothes dryer. This process produces aluminum dust particulate. Before the aluminum chips reach Furnace No. 5 for melting, the dust is blown off and collected in five filter cartridges.

Any combustible material can burn rapidly in a finely divided form. If dust is suspended in air in just the right concentration and conditions, a storm can explode. Even materials that do not burn in larger pieces, such as aluminum or iron, can explode in dust form. Dust from the production of candy, sugar, and spice can also explode into fiery infernos. Aluminum dust is ranked as one of the most explosive materials.

Typically, for dust explosions to occur, all the following conditions must be present: combustible dust is suspended or lofted in air; the dust ignites; and the dust is confined such that damaging pressures can accumulate.

At Hayes, witnesses reported that it was not uncommon to see a flare-up or bright flash above the sidewall of Furnace No. 5 whenever the chip feed was started. Investigators found that there had been a number of "near misses," or previous small dust explosions, but that Hayes management did not investigate those events, or institute training for employees to identify what constitutes a "near miss." Notwithstanding the witness accounts, Hayes had no written record of any of these fire events, including those involving fire department responses. Maybe one reason there was a lack of training and reporting is that Hayes would have been charged by the fire department for every call or "near miss." Investigators from the CSB concluded that "the fact that Hayes management had no record of—and in some cases was not aware of—these incidents and that they did not investigate them indicates lack of an effective incident recognition and investigation management system."

On the day that Shawn was killed, there was a small duct fire that, as with other such fires, untrained employees did not report to their supervisors or call in to the fire department. Employees and contractors later told CSB investigators that the fireball that killed Shawn was significantly larger than they had seen in the past, and it was orange, not the bright white flashes previously seen.

At about 2:30 p.m. on October 29, 2003, Hayes maintenance personnel shut down the chip feed system because of a smoldering fire located near Furnace 5. The glowing red duct alerted employees to the fire. On this particular day, employees followed the usual practice for dealing with these fires by shutting down the dry chip feed and the fume hood draft fan (but not the dust collector draft fan) and allowing the fire to burn itself out. The duct fire had been out for at least two hours before three Hayes maintenance personnel, including Shawn, restarted the dry chip feed system; it was about 8:20 p.m. (CSB concluded that the only likely connection between this fire and the incident later that evening was the fact that restarting the chip feed system afterward placed the maintenance employees in the vicinity of Furnace No. 5 at the time of the explosion.) Employees reported that after the restart, aluminum chips fed steadily to Furnace No. 5 for about ten minutes. At that point, one of the employees—who was standing on the north side of the furnace, about ten feet from his coworkers—noticed chips falling out of the spark box in the dust collector duct.

Maintenance personnel were familiar with this phenomenon, which typically indicated that a crust had formed, impeding chip feed, and that chips were overflowing into the dust duct. It is believed that Shawn noticed the chips falling from the spark box. As he turned to tell his coworkers to stop the dry chip feed, a fireball erupted from beneath the furnace fume hood. Shawn, who had returned to retrieve his tools before the explosion, and was closest to the furnace sidewall, was totally engulfed in flames. The fireball ignited his clothing and another mechanic standing nearby. The flames also singed a third mechanic, who was standing near the system control panel.

The fireball expanded, rose upward, and a second, more intense explosion occurred, blowing open the roof of the building. A contractor was on the roof when the explosion occurred. As recorded in CSB interviews, he heard a boom, was knocked down as the roof panels beneath his feet heaved upward, and he saw flames erupt up and through the roof. The contractor ran away from the flames, descended a ladder, and attempted to call 911. He was not seriously injured. The fire was so hot that plastic pallets containing rejected aluminum wheels waiting to be remelted were also ignited, melting the wheels.

Within minutes of the explosion, employees used the intercom system to report the emergency. The plant emergency alarm sounded, and employees evacuated the building. Coworkers assisted burn victims until emergency medical technicians arrived. A head count accounted for all employees and contractors.

The Huntington fire chief resided close to the industrial park and heard the explosion. Within five to ten minutes of the explosion he was on the scene. Because of prior visits to the plant, he knew that the Hayes plant worked with molten aluminum and understood the proper techniques and materials for fighting aluminum fires. The fire was contained and extinguished in about two hours.

Fire department emergency responders attended to Shawn and to other burn victims. The blast melted Shawn's eyes, fatally scorched his internal organs, and burned through most of his muscle tissue. Shawn did not die instantly. He lay smoldering on the floor while the aluminum dust continued to burn through his flesh and muscle tissue. The breaths that he took burned his internal organs. His last words to the EMTs before he lost consciousness were simply, "I'm in a world of hurt."

Shawn was airlifted to the burn unit at Parkview Hospital in Fort Wayne. Tammy and Mark got a call from a friend working at the plant, telling them that Shawn was badly hurt. They got in their car and raced five hours up the interstate to Fort Wayne. Shawn had fourth-degree burns over 90 percent of his body. The doctors told Tammy that they could not treat Shawn. Even if they removed his limbs, his internal organs were far too burned, evidenced by the black sludge that the doctors were pumping out of Shawn's failing body. Instead, they induced a morphine coma, slowing down his bodily functions, allowing Shawn to die peacefully. Tammy made the decision to take Shawn off life support. He died of thermal burn injuries the next day. Shawn was only thirty-three.

A second employee, David Riplinger, received serious burns over nearly half his body. He remained in critical condition for days and was hospitalized for weeks. A third injured employee received only minor localized burns and returned to work to assist investigators the day after the explosion. Four additional workers (three Hayes employees and one contractor) received treatment by emergency responders for minor injuries on the scene; none of these workers required hospitalization.

Since there were no specific regulations for combustible dust explosions, Indiana–OSHA issued citations against Hayes under the General Duty Clause, defined in Section 5 of OSHA as requiring employers to furnish a place of employment free of recognized hazards. Hayes was also cited for inadequate housekeeping and for violation of the Hazard Communication Standard, because it did not train its employees about the hazards of combustible dust. As part of a settlement agreement, Hayes agreed to suspend operation of the pneumatic conveyance systems for dry aluminum chips. Hayes also agreed to suspend similar operations at other subsidiaries. The Huntington plant has agreed to employ different methods of chip reclamation to reduce the possibility of aluminum dust explosions.

Family members grieve in different ways. Tammy channeled her grief into a lasting memorial for Shawn. She founded the United Support and Memorial for Workplace Fatalities (USMWF), a not-for-profit organization that offers support, guidance, resources, and advocacy for family members who have lost a loved one from work-related injuries or illnesses. Tammy and other USMWF members have testified before Congress about the failure of OSHA to issue new workplace regulations, and for the need to include surviving family members in the post-accident enforcement process, particularly when companies seek to abate the meager fines levied for the deaths of workers. Tammy's activism born of Shawn's death has made a difference. In April 2012, OSHA issued a new policy for communicating with next of kin.

For Shawn, he undoubtedly had plans that week to work on his small farm, preparing it for the winter. Dressed in his Carhartt jacket, work gloves, red stocking cap, and a smile, Shawn surely had plans to drag the chain saw into the woods and cut the fallen trees into uniform logs for stacking. He would put up storm windows, caulk where needed, and lay the garden to bed under a cover of hay. When Shawn went to work on October 29, he had a lot to do before the first snow fell. He had no plans to die.

Warehouse Worker

> There's no singing in the North Pole.
> GIMBEL'S MANAGER, FROM THE MOVIE *ELF*

By now, most of us know that store aisles and shelves stuffed with the latest must-have toys, clothing, and electronics for the ever-lengthening

Christmas season did not arrive there from the North Pole on a sleigh loaded by elves. For those still in the dark, large retailers like Amazon, Walmart, Target, Michaels, Home Depot, Bed Bath & Beyond, and the thousands of regional and national grocery stores are supplied by warehouse workers employed in an industry where, according to OSHA, the fatal injury rate is higher than the national average for all industries. Behind the gleaming aisles of consumer goods and groceries, there's a deadly world of crashing forklifts, falling pallets, electrical and chemical hazards, and debilitating back injuries.

For much of the year, Chicago feels like the North Pole. And like the fantastical North Pole, it is one of the most important transportation hubs in the world. Three transportation segments have made Chicago the transportation behemoth it is—sea, rail, and road. Chicago is the largest inland container port in the United States. Shipping containers stacked many stories high, most originating from China, are offloaded onto railcars and trucks destined to one of twenty-five intermodal terminals in the Chicago area with more than half a billion square feet of warehouse space. There the product is unloaded and processed by warehouse workers, who at recent count number around 150,000 in the Chicago region.[65] Replicate this in New York / New Jersey or Southern California, and other similar transportation hubs, and there are hundreds of thousands of warehouse workers running daily risks to put food on Americans' tables, clothing on their backs, and toys in their chests.

A typical modern warehouse has a dock with dozens of truck bays on one side for unloading, and another identical dock on the opposite side of the massive building for loading boxes and shrink-wrapped pallets of product onto semitrailers for delivery to retail stores. Inside the warehouse is a Rube Goldberg–like conveyor system that automatically receives the product and moves it along the highways of conveyor belts snaking up and around the warehouse ceiling to its intended truck waiting silently to be filled with its load. Other product not immediately shipped out is stored on the racking and shelving systems that reach up to the ceiling, delivered there by the dozens of forklifts racing through the narrow and congested driving lanes and warehouse aisles. Added to this deadly mix are the hundreds of "lumpers," the freight handlers, who unload the truck trailers and sort and segregate the delivery for the conveyor belt or to the forklift driver. Separating the lumper from the speeding forklifts or belt line is usually nothing more than a yellow line on the warehouse floor, a painted "curb" between where the lumpers and forklift drivers work. Because most warehouse workers are paid a piece rate based on how much product they can unload, load, and store, speed is at a premium, because it determines how much money they can make. Speed is a known hazard, yet workers can be disciplined if they do not work fast enough. And therein lies the danger.

Speed kills and injures. Speed kills and injures as a worker drives from bay to bay in order to make rate. This can cause tip-overs, crashes, and resulting death and injury from falling pallets crushing the driver or nearby workers. Speed kills when a forklift driver in a hurry stacks a rack or shelf with pallets of bottled water that collapse onto workers below. And speed injures when a lumper blows out a disc in a hurry to pick up a box that exceeds the weight that he or she can safely lift without assistance.

In Southern California's "Inland Empire," east of Los Angeles, more than one hundred thousand workers toil in the massive warehouses there. The workforce is mostly Latino and mostly immigrant. They are often temporary workers, who are paid low wages, and without benefits. According to a study of these workers, the rate of injury for temporary workers in the Inland Empire was more than ten times greater than the national rates for warehouse industries.[66]

Unlike the elves in the North Pole, there is no benevolent and gentle Santa and Mrs. Claus to oversee these workers and to give them Christmas cookies and candies for snacks. For these workers, there's no singing in the Inland Empire.

Yvonne Shurelds

She was a small woman, only five feet tall. But she had a big and infectious personality that could fill up a room. Her face was broad and handsome, as was her smile and smiling eyes. Yvonne Shurelds lived her entire life in Lima, Ohio, tucked comfortably into the northwest corner of the state. While family and friends dispersed, drawn by the lure of big cities and warmer climes, Yvonne happily stayed put in the safety and security of her hometown.

Lima became known for its industrial manufacturing by the middle of the nineteenth century. Later, oil was discovered there in 1885, and the World Wars established Lima as a manufacturing center for trucks and other vehicles, including the Pershing tank. Today, Lima and Allen County are home to thirty-three automobile-related manufacturing businesses employing more than five thousand people. Lima was a good place to work and live.

Born on Christmas Day in 1949, Yvonne was the third of eleven children—seven girls and four boys. Helen and "L.V." Shurelds were from around Tupelo, Mississippi. After retiring from the army, L.V. migrated north to work for Standard Oil in Lima. Helen found work as a nurse at the Lima Memorial Hospital. They were a big, close family. Helen was the glue. After a while, L.V. got in a horrific car accident and was forced to retire young because of his injuries. He busied himself cooking for his family, friends, and their church. L.V. was a wonderful cook, specializing in Creole soul food. Étoufées, catfish, red beans, and rice.

Yvonne helped care for her siblings, especially her youngest brother, Bruce, known as "Baby." At Lima Senior High School, Yvonne had big dreams of going away to college, but she got pregnant, and her life and dreams changed forever. Yvonne became a single parent and stayed back in Lima. She enjoyed being a mother to a son and daughter, and began a long career working for RCA at its semiconductor plant in nearby Findlay. She worked at RCA for thirty-two years, until the plant closed in 2002. Yvonne had a good pension and spent time in her retirement traveling with Helen and shopping. Yvonne loved bargain shopping at thrift stores and estate sales. She restored antique furniture and collected elephants. And she loved her dog Behr, a white poodle that she called her second son. Yvonne's son and daughter, Marquette and Wendy, both live in California. They tried to persuade her to move there. But Yvonne would not leave Helen. They were together every day, especially after L.V. died.

After a couple of years into her forced retirement, Yvonne wanted to go back to work. She loved to work. She loved to meet people and had lots of friends from her years at RCA. In 2004, Yvonne was hired on at DTR Industries in Bluffton, Ohio. DTR is a Japanese-owned company that manufactures automotive rubber parts, including for Toyota. She initially worked on the assembly line, and later transferred to shipping and receiving. There, Yvonne drove a cart around the plant, picking up tools and parts. This gave her the opportunity to travel around and socialize with her coworkers. She was well liked. Everyone knew Yvonne at the plant.

In the first week of March 2008, her supervisor assigned Yvonne to cover for a coworker while he was on vacation. The coworker drove a stand-up forklift. Yvonne had never been trained to drive a forklift. Most stand-up forklifts are designed the same: The operator stands in a cab to drive the forklift. The forks and their hydraulics are in front. The back of the cab is open, generally with no corner posts to guard and protect the operator from injury when traveling backward.

The most common injury to operators of stand-up forklifts is from the crossbeam of a shelf in a warehouse entering the cab when the forklift is being driven in reverse. This is known as an "under-ride accident." Essentially what happens is this: A typical warehouse has aisle after aisle of shelving—called a rack system—that holds pallets of product moved there for storage by the forklifts. A heavy metal crossbeam runs horizontally underneath the shelving for support. If the forklift backs into the crossbeam, an under-ride will occur.

When Yvonne was assigned to cover for her vacationing coworker and to operate a stand-up forklift, she complained that she had no training or experience. Yvonne told her supervisor that she did not feel competent to operate the forklift, and that the forklift was difficult for her to maneuver. Her supervisor pushed back and claimed that Yvonne was taking too long to learn how to operate the forklift. On Friday, March 7, Yvonne called her daughter Wendy, who lived in San Diego, and said that she was afraid to operate the forklift. Yvonne was killed the next day.

On March 8, 2008, Yvonne was scheduled to receive additional training on operating the forklift. However, a late-season snowstorm hit the Lima area, and most workers did not report for work, including Yvonne's trainer. Yvonne, however, showed up to work as scheduled. And without proper training, she was assigned to pick pallets stored on the shelves. She worked the second shift alone that day in the warehouse. When they found Yvonne, she was lying on the floor unresponsive, blood oozing from her nose. The crossbeam had crushed her head. The official report was internal decapitation, which occurs when the skull separates from the spinal cord.

Yvonne's nephew Chris was the first family member notified of Yvonne's accident. Chris called Wendy and Marquette in California. He knew only that there was an accident involving Yvonne. No one would tell him, or her children when they called, anything more. The alarm was sounded, and other family members, including Helen, came to the plant in a blizzard. Helen and Chris were told that Yvonne had died, but Helen was not allowed to identify the body. A coworker identified Yvonne's crushed body instead.

Without her family's knowledge or permission, Yvonne's body was shipped to Toledo for an autopsy. When Wendy arrived from California the next day, she still did not know how Yvonne had died, or where her body was. Later, Wendy learned that her mother was in Toledo. Wendy traveled there to recover Yvonne's body.

Yvonne's funeral was delayed until March 18, when all her friends and family could travel back to Lima. Over five hundred persons attended her funeral. It was held at the largest church in Lima.

Yvonne's death did not have to happen. Simple solutions would have avoided her death. First, the crossbeams, or the shelves on the rack, should have been set at a height that would have prevented an under-ride. If the shelves were set at the same height as the forklift's overhead guard, the crossbeam or shelf would have hit the guard, thereby preventing the crossbeam from entering the operator's cab. A second simple design solution would have been to provide rear vertical posts preventing the crossbeam from entering the cab in the first place. Without these design changes, even an experienced operator is at risk.

OSHA investigated Yvonne's death and issued a citation and notification of penalty. Seemingly unrelated to Yvonne's death, the citation found four instances that DTR did not provide truck operators with protective equipment when changing and charging lead acid batteries, thereby exposing operators to corrosive acid splashes. The proposed penalty for each of these violations was $1,800. Related to Yvonne's death, OSHA cited DTR for failing to provide sufficient training on how to operate the forklift truck related to "maneuverability" and utilizing a control stick for "acceleration and deceleration," as well as how to operate the steering wheel. Basically, OSHA found that DTR did not train Yvonne in how to drive the forklift. For this violation, which caused her death, OSHA proposed a penalty of $4,500. Two other citation items were related to the forklift not being in safe operating condition and not being examined before being placed in service. The proposed penalty for these violations was $2,500. In all, the proposed penalty related to Yvonne's death totaled $14,200.

According to Dun & Bradstreet, DTR had reported annual revenues of over $65 million. OSHA regulations provide that an employer may contest and seek an abatement of a proposed penalty. DTR did just that. On June 10, 2008, three months after Yvonne died of internal decapitation, OSHA and DTR entered into an Informal Settlement Agreement that reduced the proposed penalty by half, down to $7,100. The last paragraph of the agreement states in airtight legalese that nothing in the Agreement "shall be deemed an admission by Respondent of the allegations contained within the citations at issue in this proceeding." It goes on for several more lines, including that the Agreement was "made solely for the purpose of settling this matter economically and amicably." In plain speak, DTR took no responsibility for Yvonne's death and only begrudgingly paid the fines.

There is no doubt that DTR settled Yvonne's death cheaply; DTR did not have to dig deep into its revenues to pay this measly penalty. Still, two things stand out in this Agreement, which in truth are not all that unusual. First, DTR took no responsibility for Yvonne's death. The Agreement says so. DTR settled on the cheap to rid itself of the nuisance of Yvonne's death. Second, Yvonne's family was not even given the courtesy of being consulted before OSHA conducted a fire sale by giving DTR a free pass on her death. OSHA allowed DTR to walk away without any acknowledgment that the citation findings had any validity and that Yvonne's death was a preventable event.

For a family that lost a daughter, mother, sister, grandmother, and aunt, and for the hundreds of friends who flocked the church for her funeral service, the Informal Settlement Agreement was a cold slap in the face. It prevented closure, keeping the wound of her death open and, in the parlance of OSHA, unabated.

With their legal options for answers and accountability forever foreclosed by the OSHA agreement, and DTR's immunity from intentional tort liability under Ohio workers' compensation law, Yvonne's survivors struggled to make sense of her death. For Wendy, she found an outlet by becoming an advocate for workers' rights. She spoke about her mother's senseless death at Southern California Coalition for Occupational Safety and Health events, and before Congress on the importance of family involvement in OSHA's investigation and enforcement process. Her brother was not as fortunate. He suffered emotionally, along with Helen, whose health declined after Yvonne's death.

In the end, OSHA's findings were not intended to bring Yvonne back. No one had any such illusions. But, in a small way—$7,100 small—the findings and citations were supposed to act as a deterrent against the next employer who puts a worker in harm's way. However, with deterrence unavailing, it is just a matter of time, likely moments, before the next worker dies a preventable death.

Packinghouse Worker

> They had chains which they fastened about the leg of the nearest hog, and the other end of the chain they hooked into one of the rings upon the wheel. So, as the wheel turned, a hog was suddenly jerked off his feet and borne aloft. At the same instant the ear was assailed by a most terrifying shriek; the visitors started in alarm, the women turned pale and shrank back. The shriek was followed by another, louder and yet more agonizing—for once started upon that journey, the hog never came back; at the top of the wheel he was shunted off upon a trolley and went sailing down the room. And meantime another was swung up, and then another, and another, until there was a double line of them, each dangling by a foot and kicking in frenzy—and squealing.
>
> UPTON SINCLAIR, *THE JUNGLE*

Not much has really changed in the way hogs are slaughtered in a modern-day packinghouse since 1906 when Upton Sinclair wrote *The Jungle*. The same can be said for beef and poultry. Stepping inside a slaughter and processing plant today is, without any exaggeration, a hellish vision. A visit to a packinghouse is not for the faint of heart, or stomach.

The animals are offloaded into pens from semitrailers and crates and funneled into the mouth of a packinghouse where they are stunned and rendered unconscious. On the next stop along the line, the animals are shackled and hung. Throats and bellies are split, bleeding the animals to their final death, while workers are elbow-deep inside the large animal cavities eviscerating and removing intestines and internal organs. Not much has changed. There is a smell, not necessarily revolting, but unlike anything you have ever smelled, unless, of course, you have lived downwind from a packinghouse. It is an odorous mixture of raw meat, blood, hides, feathers, feces, dirt, body odor, renderings, smoke, and of death. The fear sensed by thousands of animals herded down a narrow chute to their deaths produces a smell, according to animal scientists. Chickens are known to scream in a frenzy of fear, often defecating and vomiting on each other and on the workers. Then there is the noise, a cacophonous symphony of squeals and braying, screaming birds, shackles, electric saws, head splitters, conveyors, powered hoses, beeping forklifts, and a babel of voices from hundreds of workers. Spanish, English, Burmese, and Somali are the most common languages in a packinghouse today. A packinghouse is a wet place, slick with blood, water, chemicals, steam, and sweat. Finally, there are wide temperature extremes inside a plant: cold for storage and processing, and hot for cooking and rendering. And, above all, a slaughter and processing plant is an extremely dangerous workplace for the men and women working there.

All of the slaughter and workplace hazards occur to satisfy the insatiable appetites of carnivores around the globe. In 2015, Americans alone consumed per capita almost 211 pounds of red meat and poultry. Annually, this adds up to more than 93 billion pounds of meat and poultry produced in the United States for feeding 318 million Americans, according to the North American Meat Institute.

For workers toiling in slaughter and processing plants in the United States, not much has changed since 1906 when it comes to their health and safety. While overall workplace illness and injury rates have declined, and not necessarily because the plants and jobs are safer, meat and poultry workers still have one of the highest injury and illness rates of any industry. Food manufacturing workers are twice as likely to suffer injuries and illnesses as workers in the industrial and manufacturing sectors. In meat plants, according to the BLS, the rate of injury and illness per one hundred full-time workers in 2014 was 8.7, compared to 3.3 for workers overall.

Although mechanization is the norm in a modern processing plant, many jobs are still done by hand with sharp knives and saws, including evisceration and cutting. As a result, most injuries are from cuts, lacerations, punctures, and musculoskeletal disorders (MSDs) caused by the trauma of repetitive cutting motions. The line at the plant nurse for ice and ibuprofen is long and a necessary stop for many workers in order to get through the workday. Other common injuries are from falls on the slippery floors, an accident made all the more dangerous when standing in a line of workers holding boning knives. The smell in many plants lingers over the workplace like a toxic cloud, owing to poor ventilation, which contributes to respiratory illnesses. Workers are at risk from extreme temperatures, including for hypothermia and frostbite in the freezer rooms, and heat stress and dehydration in the oven rooms. Animals can expose workers to bacteria and pathogens from blood and fecal sources, including salmonella. It is not uncommon for a cow to be hung before it finally dies, making workers nearby at risk from being kicked by a wildly thrashing thirteen-hundred-pound animal.

In poultry plants, workers must endure dangerous line speeds of more than 100 birds per minute, which is the number of birds that a group of workers hang, gut, or slice in one minute. The poultry industry lobbied hard to increase the maximum allowed line speed from 140 birds per minute to 175 per minute, despite the overwhelming evidence that the current line speeds were injuring poultry workers at unacceptable rates. Poultry workers suffer serious injuries at a rate that is almost double the rate of private industry. MSD injuries, particularly carpal tunnel syndrome, is more than seven times the national average. Like workers in meat plants, poultry workers suffer from MSDs as they make repetitive cuts in order to bring us a boneless chicken breast, or chicken fingers for our kids. In March 2014, NIOSH reported, among other findings, that 42 percent of poultry workers have evidence of carpal tunnel syndrome, while 57 percent reported at least one musculoskeletal symptom. NIOSH researchers found that 76 percent of poultry workers had abnormal nerve conduction test results in at least one hand.[67] Workers suffering MSDs to their hands and wrists reported difficulty buttoning their shirts, holding a spoon, or picking up a glass of water. Recommendations for reducing these injuries included allowing more than one break per shift, requiring more frequent job rotations, and using sharper knives to reduce the stress from cutting. To prevent workers from self-medicating (because company nurses were overwhelmed), NIOSH even recommended the removal of a medicine dispenser in the cafeteria at one plant that dispenses pain medication. In addition to MSDs, cuts and gashes are also common, as workers stand shoulder to shoulder wielding knives while the birds speed by for processing. The poultry industry lobbied to increase line speeds as part of a larger USDA food safety rule, and in the end a coalition of food safety and worker safety advocates prevailed in opposing the line speed increase. But it was a pyrrhic victory, as the working conditions of poultry workers are unsafe at any speed. In fact, there is no OSHA rule that regulates line speeds. In March 2015, OSHA denied a rule-making petition to regulate line speeds at poultry plants, citing its "limited resources," while at the same time acknowledging that the incidence of occupational injuries to both meat and poultry workers remained high.

Although it may be an unappetizing thought before taking that first juicy bite of a sizzling steak, it behooves us to remember the workers who slaughtered and processed the cow so that we can enjoy a bone-in filet, porterhouse, or T-bone. The filet slightly charred, or the succulent chicken marsala, were made possible by workers who may no longer have the hand strength to cut their own dinners.

Eufracia Barrera

Morelia sits in Mexico's central highlands around four hours northwest of Mexico City. It is the capital of the state of Michoacán and home to beautiful colonial churches and haciendas dating to the sixteenth century. UNESCO has designated Morelia as a World Heritage Site because of its natural and cultural beauty, including as many as one billion monarch butterflies that migrate there each year, more than twenty-five hundred miles from eastern Canada.

In a different kind of migration, Eufracia Barrera traveled with her three children in 1995 from Morelia to Monticello, Indiana. Monticello is the county seat of White County, located about halfway between Chicago and Indianapolis and carved out of surrounding farm fields. Its motto is "Life with a Splash," because of its proximity to Lakes Freeman and Schaefer, which provide the reason for outsiders to visit the area in the summer. Beyond the twin lakes, Monticello is a typical small midwestern town, with around fifty-seven hundred residents. In 2005 a fire destroyed one of the last remaining manufacturing plants, leaving four city blocks polluted with toxins as its legacy. The economic base is now made up of local small businesses, fast food chains, and a Walmart.

Eufracia is a tiny woman, maybe five feet tall in her shoes. She has a big smile, especially when she talks about her children, now all grown and working. She followed her husband, Mateo, who arrived in Monticello two years earlier looking for work. He found it as a punch press operator for a company that made small motors for refrigerators at a plant that later closed. "Years for us went by in barrels," she said. "It was hard to be apart at age thirty-eight with three high-school-age children," two girls and a boy. Their children were the first Hispanic kids at the high school. Now there are many more. In the 2010 census, Hispanics or Latinos made up 12.5 percent of Monticello's population.

By the time Eufracia arrived in Monticello, there were already other family members in town. Her sisters Amada and Marguerita and their husbands and children came to Monticello via Chicago, not just for the work, "but because it was calm here for our children," Eufracia said. "We all gave our children a good foundation."

Amada and brother Armando were working at the Tyson Foods pork plant in Logansport, around twenty miles east of Monticello. Tyson Foods had $33 billion in sales in 2012, operating more than one hundred production plants in the United States. Known mostly for chicken, Tyson also operates nine pork production plants, slaughtering and processing more than four hundred thousand hogs a week. It is an industry brag that every part of the hog is slaughtered and processed except the oink. Eufracia wanted to work at Tyson, and she asked if Armando would give her a recommendation. At the time, Tyson was paying workers $1,000 for a referral, but Armando still said no to the recommendation. He said the work at Tyson was too hard for Eufracia. Armando jokingly said he would give Eufracia the referral fee, and then she could quit. Eufracia ignored Armando's advice. "If there were other ladies working there, I had to try it," she said.

At the Logansport plant, Tyson has increased its line speeds, the speed at which the hogs are killed and processed. Tyson now processes over a thousand hogs an hour, more than when Eufracia first started, although the number of employees has remained steady at around seventeen hundred production workers. The more hogs that can be killed and processed by the same number of workers, the more profit goes to the corporate bottom line. "We just have to work faster with the same number of workers."

Eufracia started working in "guts" and later moved to "knuckle meat." It is not actually the knuckle of the hog's feet, but the extreme shank end of the leg bone just below the ham, also known as the ham hock. Workers use a hook to yank and cut off the knuckle from the ham, filling large containers with the meat. After three years doing this job, Eufracia bid on the skinner position, skinning the knuckles of excess fat. This involved flipping and rolling and turning the knuckle in a drumlike gear that pulled the meat toward a blade that skinned the fat. It was difficult work, requiring the worker to use a lot of force while manipulating the knuckle around. It put an inordinate amount of pressure on the arms and shoulders for long periods of time. The same motions are repeated thousands of times a week, too many times for Eufracia to even begin to count. Ergonomic experts have reported that it is not uncommon for a line worker in a slaughterhouse to make forty thousand repetitive cuts in a single shift. Eufracia worked seven years skinning fat from the knuckle, nine hours a day, six days a week. There were a lot of hogs to process.

In October 2010, Eufracia felt a sharp pain in her right shoulder while skinning a knuckle. She reported it to her supervisor. He sent her to the plant nurse for ice and ibuprofen. There, Eufracia and several other workers sat with bags of ice on their shoulders, wrists, elbows, fingers, necks, backs, and other assorted body parts and joints. Fifteen minutes of ice and ibuprofen, and then back to the line without any X-ray, or proper shoulder exam, to determine what was wrong. Instead, the nurses merely asked the same simple questions: "How do you feel?" "You think you can work?" "Even with your pain, you got to try it." Eufracia would always answer the same. "It hurts." On a scale of one to ten, "I always said ten, and the nurses would write down a seven." When their fifteen minutes were up, other workers took their place. All day long there was a constant stream of workers going in and out of the nurse's office. Ice, ibuprofen, and back to work. At the end of the workweek, Eufracia was sent home with a handful of ibuprofen to get her through the weekend. Tyson insists that the employees first try this "regimen" before it will send them to the company doctor.

Repetitive injury in the meatpacking industry is double the injury rate of two dozen other high-hazard industries identified by OSHA, such as construction, mining, and shipbuilding. These injuries often leave the workers irreparably damaged and disabled. It is not uncommon for meatpacking workers to have to run their hands under hot water when they wake in the morning just so they can open their fingers frozen in an arthritic claw.

Eufracia followed the plant's treatment regimen for two more months, sometimes making multiple daily visits to the nurse. In December, she reported that her right shoulder felt like it was dislocated. Still, she was treated only with ice and ibuprofen every four hours. But her range of motion was limited. Eufracia's shoulder made popping noises if she tried to raise her right arm above her waist.

By March 2011, Eufracia could no longer tolerate the pain, which she always registered as a ten, when asked. More than four months after she first reported the injury to her shoulder, Tyson relented and approved a doctor's visit. Not surprisingly, an MRI showed that all the tendons in her right shoulder were torn from the bone. The official diagnosis was a torn rotator cuff, an injury commonly associated with a baseball pitcher, but far more common for meatpacking workers. There are four rotator cuff tendons that surround the ball of the shoulder joint. The tendons control the rotation of the shoulder and keep the ball of the joint centered. When Eufracia reported in December that her shoulder felt like it was dislocated, it felt that way because it was. What probably started as a small tear in October, treatable with physical therapy and rest, turned into a full-blown four-tendon tear two months later, as a result of the ice, ibuprofen, work, and neglect by Tyson for the well-being of Eufracia.

In April, Eufracia had surgery to reattach the tendons to her shoulder. It can take up to twelve weeks for the tendons to reattach to the bone in the best of circumstances, including no reaching or lifting movement to the shoulder. In these best of circumstances, it takes most patients six months to feel pain-free, and twelve months to regain their shoulder strength. However, Tyson's doctor returned Eufracia to light-duty work two weeks after surgery, and after only one week of physical therapy at the plant. Light duty at a pork processing plant is not exactly sedentary, like filing or answering the telephone. Eufracia was back on the disassembly lines, this time pulling lard and fat from cut pieces of pork with one hand. That was light duty at Tyson. For three months, Eufracia did this work left-handed, and on one of the fastest lines at the plant. At home she wore her arm in a sling, but was not allowed to do so at work. Instead of immobilizing her right arm and shoulder in a sling, as is medical protocol, she stood for hours while her right arm dangled by her side, swinging to the movement of her left arm pulling fat and lard. After three months, Tyson's doctor released Eufracia to full duty. Still, because her shoulder had not properly healed, she could work only one hour before the pain became intolerable. But she continued to work because she had a full release, and not working could cost her her job.

After more than three months of using only her left arm, she began to experience shooting pains in her left hand, and then her left arm would fall asleep. She was not warned that pulling fat and lard one-handed could overwork and damage her good left arm. After more ice and ibuprofen, she was eventually sent back to the company doctor. Again, no surprise, the MRI showed torn tendons in her left shoulder. But this time, with the workers' compensation adjuster sitting in the examining room, the doctor said Eufracia had reached her "MMI," maximum medical improvement, meaning that there was nothing more he could do for her. Eufracia was broken down beyond repair and could not work anymore. Devastated by this news, she asked for her X-rays. She wanted a second opinion. The doctor threw the records at her and yelled, "Can you read English?"

In Indiana, like many states, the company, or rather their workers' compensation insurance carrier, picks the treating doctor. Workers are free to go to their own doctor, but insurance won't pay. An obvious conflict of interest between doctor and patient is created by this system of insurance and treatment, with the insurance company in the middle calling the medical shots. It goes this way: In Eufracia's case—which was typical—when she was finally referred to a doctor after months of ice and ibuprofen, it was via a referral made by the workers' compensation insurance adjuster. The doctor is happy to receive referral business, especially from a big insurance carrier representing a big company like Tyson, with lots of injured workers. Doctors can make a nice living from the referrals, but only if they don't upset the insurance adjuster by being a patient-friendly doctor. The less treatment a doctor provides, the lower the permanent partial impairment (PPI) rating the doctor assigns to the permanence of the injury, the less money the insurance company has to pay, and the more business the doctor is referred. All is good—except for the injured and disabled workers.

Eufracia said she is too damaged to find other work. No one would hire her. At her own expense, Eufracia went to another doctor. He gave her steroid injections in her neck, but she still had torn tendons and constant pain. She can't do simple tasks, like opening jars, without pain and popping in her shoulder. She can't sleep at night, or hold her grandchildren. She is depressed.

Sitting in her modest and neat living room, with a crucifix looking like it was made from popsicle sticks hanging on the front door, Eufracia was in tears. She was alternating between anger at Tyson, the doctor, the insurance company, and the fear of living in pain for the rest of her life, too injured to do anything without pain. Yet, she persists. "I'm not lazy.

I'm a hard worker. I want to work," she said. The doctor sent her back to work before she was ready because the insurance company didn't want to pay out benefits for her to stay at home and heal. Now she pulls fat and lard with her right hand, which has more strength than her left. She holds up her left hand for me to see. It is shaking. She cannot control it. Her shoulder is too weak to support her arm. Eufracia knows that she is at the end of the line. She is hoping for a settlement with the insurance company, but knows that it won't come close to compensating her for the permanency and pain of her injuries, not to mention that she will never work again. In October 2013, Indiana decreased the cost of workers' compensation insurance paid by employers by 7 percent, a trend that began several years earlier. Indiana workers' compensation insurance rates are 43 percent below the national median. Only North Dakota has lower workers' compensation rates.

On her couch, her arms carefully cradled in her lap, tears drying, Eufracia looked up and smiled. "I should have listened to Armando." She then suggested the title of her story. "Como dejamos nuestras armas en el trabajo," Eufracia said: How we leave our arms at work.

Manufacturing

Modern Times, Charlie Chaplin's last silent film, and final bow for the Tramp, is a none-too-subtle excoriation of the excesses that industrial capitalism unleashed on American workers in the Great Depression. Its most iconic scene, where the Tramp is caught in the wheels and gears of a great machine, has been acclaimed as a brilliant visual metaphor for the dehumanizing effects of the modern factory system, where factory work will grind up and spit out workers. However, for American factory workers today, being ground up and spit out is not a metaphor, but a dangerous reality, and is an all too frequent occurrence.

The National Association of Manufacturers boasts that manufacturing supports an estimated 18.5 million jobs in the United States—about one in six private-sector jobs—and that more than twelve million Americans (9 percent of the workforce) are directly employed in manufacturing.[68] According to the North American Industry Classification System (the standard used by the federal government to classify businesses for the purpose of collecting and analyzing economic data), manufacturing is often described as plants, factories, or mills and characteristically uses power-driven machines and materials-handling equipment, and those in manufacturing are "engaged in the mechanical, physical, or chemical transformation of materials, substances, or components into new products."[69] It is a diverse sector that is further defined by its various manufacturing processes. These processes include casting, molding, forming, machining, and joining. All the processes require the use of power machinery and tools by a machine operator, who may roll thick plate and sheet metal; shear, pierce, and stamp metal, leather, or other materials; or weld, turn, lathe, drill, shape, plane, and finish the manufactured product.

These manufacturing processes and job functions are accompanied by mechanical hazards, some known and guarded, while others occur without warning. OSHA's publication on the "Basics of Machine Safeguarding" describes the hazards as follows: "Crushed hands and arms, severed fingers, blindness—the list of possible machinery-related injuries is as long as it is horrifying. There seem to be as many hazards created by moving machine parts as there are types of machines."[70]

There are three basic danger zones that require safeguarding in some fashion, because of dangerous moving parts in all machines. The point of operation is where the work is performed on the material. It can involve cutting, sawing, shaping, boring, or forming. Different mechanical motions and actions present hazards to a worker from rotating parts, feed mechanisms, meshing gears, belts, saws, shearing, punch presses, and cutting teeth, among many more. Basically, any moving part is a hazard. To prevent an injury or death, safety rule number one is to prevent contact between the worker and the wrong part of the machine. It is easier said than done, since workers have to be in close contact to operate and maintain their machines.

Guards are the preferred method of protection. Guards are physical barriers that enclose dangerous machine parts and prevent employee contact with them. Guards also protect against the introduction of objects into the machine, such as small tools, which if they fall into a machine can be rocketed back out, severely injuring a worker. The second preferred method is the use of safeguarding devices that prevent inadvertent access by employees to hazardous machine areas. This might be an automatic power cutoff if a worker opens a gate or small door that is an access point to the moving parts of a machine. Even if appropriate guards are in place, maintenance and training are other critical components to a safe workplace. If a machine is not properly maintained, it can become a danger when, for example, electrical wiring and systems are frayed and exposed, creating a hazard for electrocution. Workers must be trained in lock-out / tag-out procedures, to prevent the sudden energizing of a machine during maintenance.

Preventable manufacturing injuries and fatalities occur every day. According to the BLS, in 2014 there were 345 fatalities among workers engaged in manufacturing processes with mechanical machines, a 9 percent increase over the previous year. Add to this the fact that the total recordable injury rate in 2014 was 4.0, higher than the rate for all workers.[71] Lack of guards, training, and procedures are almost always present in post-incident root-cause analyses.

In addition to injuries and fatalities from machines, there are nonmechanical hazards. Noise is a significant cause of occupational injury. According to BLS, 12 percent of all reported occupational illnesses in 2010 for all workers were due to significant hearing loss, or around eighteen thousand workers.[72] The CDC reported that in 2007, 82 percent of occupational hearing loss occurred in manufacturing.[73]

The image of Chaplin's Tramp, tightening huge bolts while lying inside the great geared wheel, may have been seen by many moviegoers as comic. But it was not meant to be funny. It was a visual indictment of how modern factory workers are treated. For Chaplin, workers were cogs in the machinery of America's industrial output. Today, not much has changed. And when it comes to their health and safety, they are machine parts, easily replaceable when broken.

Carlos Zetino Chavez

Since his workplace injury on October 28, 2002, Carlos Zetino Chavez spends most of his days confined to a small two-bedroom apartment he shares with his wife, Sonja, in Wyoming, Michigan, a suburb of around seventy-two thousand persons, just south of Grand Rapids. The apartment is clean and comfortable, with a big flat-screen TV to keep Carlos connected to the outside world and to his beloved *fútbol* games. Everywhere you look, there are images of Jesus and the cross, including a small shrine in the corner of the living room with a red ceramic Christ standing smiling, palms outstretched.

Wyoming, established in 1832, is a far cry from La Maquina in southwest Guatemala, a small town in between the capital of Guatemala City and the Pacific Ocean. That is where Carlos lived until 1986, when he followed his sister and brothers to Los Angeles and the promise of opportunity that has called hundreds of millions of immigrants to America. Carlos prefers the climate of western Michigan to the heat and humidity of Guatemala. He never walked in snow until Wyoming. There is a vibrant Hispanic community in Wyoming. Hispanics make up the second-largest demographic there, attracted to Wyoming just as Carlos and his brothers and sister were by a solid manufacturing base for auto parts, food processing, electronics, heavy machinery, and even fire engines. There are lots of jobs that American workers will not do. Wyoming is also home to Carlos's church, San José Obrero—St. Joseph the Worker.

Carlos has always worked hard. He has always looked for the next better job to support his young family of three boys. In Guatemala he had a good job working for the government in something related to transportation and development. He remembers a Chinese economic mission teaching local Guatemalans how to grow better rice. But political change brought new managers and uncertainty to his work. Carlos figured that was as good a time as any to heed his siblings' calls for him to come *norte* to Los Angeles. The call was like a mantra. "America, America, America," Carlos said. He was thirty-two, and his second son, also Carlos, was just born. Leaving his family was difficult, but he felt it was necessary for everyone's future. So he paid a *coyote* $1,100 quetzal, the equivalent of around US$1,000, to help smuggle him into the United States. A bus to Mexico City, another to the Arizona border, and a final bus waiting for their group on the other side to take them to Los Angeles. It took a week, and in 1986 it was not as difficult as it is today. Not long after, Carlos became a U.S. citizen, part of Ronald Reagan's amnesty and immigration reform. Carlos explained again that it was different back then.

Arriving in Los Angeles, Carlos moved in with his sister Counsuelo, in her apartment on Santa Monica Boulevard in Hollywood, near Griffith Park. Eventually he shared an apartment in the same building with some friends from Guatemala. In Los Angeles, he quickly secured work as a day laborer doing construction. A couple of weeks later he got a job working as a dishwasher and in kitchen prep at the Rose Tattoo, a high-end restaurant frequented by actors and artists; Carlos met Sylvester Stallone there. Carlos worked there until sometime in 1997 or 1998, when he moved on to Michigan. Carlos has a poor memory for dates. Remembering has been a challenge since his injury.

His youngest brother, Luis, was the first to leave Los Angeles for Michigan. He found a good-paying job at the Michigan Wire Processing Company, a high-quality wire processing company for critical and commercial applications, including air bag components, conveyor components, ball and needle bearings, drivetrain and engine components, fuel injectors, high-grade fasteners, roller bearings, suspension parts, and track pins. Luis reported that there was work for Carlos at Michigan Wire, and, seeking still better opportunities, Carlos headed east for Wyoming. At Michigan Wire he eventually became an operator, cutting metal rods into pieces according to the customer's specifications. Carlos was working at a "cold drawing" process when he was injured. Cold drawing is used to reduce the cross-section area as rod or wire is pulled through a drawing die, which not only decreases its diameter but also improves surface finish and dimensional tolerances and changes its hardness, tensile strength, and shape. Although a round shape is most common, steel wire can be drawn into many other cross-sectional geometries, including ovals, hexes, half-rounds, and squares.

Carlos enjoyed the work at Michigan Wire. He made minimum wage but worked forty hours a week and overtime on occasion. He also had health insurance. The owner, Carlos said, was a good man. He treated his workers well. Carlos got paid weekly, and sent most of his paycheck back home to Sonja and his boys. When he could, Carlos traveled home to La Maquina once or twice a year, he said. His son Carlos remembers it as less frequently than that. Still, he went when he could.

Carlos settled in to Wyoming and his work at Michigan Wire. He lived with his other brother, Moises. In his spare time he hosted a radio show on Radio Exitos, WYGR 1530 AM, a Hispanic radio station in Wyoming / Grand Rapids. Carlos played *musica romantica* every day after work. He was known to his audience as "Bon Bon" because of the opening line to his show: "Open your minds and hearts to the sweet candy of music." Carlos played romantic music as Bon Bon until his injury. His favorite song was "Recuerdo de amor"—Memories of Love. Radio Exitos shut down sometime after his injury.

Carlos's other love, *fútbol*, kept him busy when he was not at work or on the air. He started the first Hispanic *fútbol* league in Grand Rapids. Carlos was a goalkeeper. Now he can only watch his favorite team, Brazil, from the living room.

On the day of his injury, Carlos was not scheduled to work. He had worked the day before, Sunday. But on Monday, a coworker did not show up for his shift. So Carlos volunteered to work on his day off. He started as usual at 6 a.m. Beyond that, Carlos has no memory of the day. He knows only what others have told him.

Carlos operated a large machine that fed steel wire as thick as his forearm onto a spool. The spools were so large that only two or three could fit into the back of a semitrailer. Carlos had to carefully control the speed of the machine so that at the end of the spool the steel wire would not whip out of control into an operator's leg or arm. Sometime during his shift on that Monday, one of the chains that held the machine to the floor broke and snapped back, crashing through Carlos's hard hat and breaking his skull.

Carlos was in a coma for twenty days. The company called Counsuelo, Carlos's emergency contact, and reported the accident. Counsuelo sent out the alarm to Luis and Moises, as well as to the cousins living in the area. Back in La Maquina, there was no telephone at home. A telephone call reporting the accident was made to a relative living in a nearby larger city. Sonja had to travel to the telephone to learn of Carlos's injuries. Michigan Wire contacted the U.S. consul in Guatemala and got Sonja an emergency visa and an airplane ticket to Michigan. Two days later, Sonja was on her first airplane, flying to Carlos. It took Carlos's minor boys longer to get to Michigan because they needed both the signature of their mother and father to obtain a Guatemalan passport. Luis had to fly back to Guatemala with proof of the injury before he could sign for their passports. The family was finally reunited on November 13. Except for a few brief trips back to Guatemala, the family has stayed in Wyoming ever since to care for Carlos. They all live across from each other in the same apartment complex. His oldest son, Selven, is now married, and with a grandson for Carlos.

Carlos spent four months in intensive care at Spectrum Hospital in Grand Rapids, and another six months in a hospital room before he began three months of grueling therapy, learning to walk and talk all over again. Carlos has had five surgeries on his head. He does not know how much his medical expenses have been. As part of his workers' compensation benefits, his medical bills have all been paid.

Today, looking at Carlos, it appears as if he had suffered a severe stroke. His speech is strained and slurred. He keeps a washcloth on his lap to wipe the drool from his mouth. The left side of his body does not work any longer. He wears a brace around his left foot and ankle because he cannot support it otherwise. Carlos's left hand is a useless claw. And he is losing his vision.

For all his pain and suffering, Carlos was limited by law to workers' compensation benefits. Michigan has had workers' compensation laws since 1912. *The Michigan Business Guide to Workers' Compensation*, ninth edition, explains that the purpose of workers' compensation is two-fold: to provide benefits to workers injured on the job and "to protect employers from costly litigation over claims of work-related injuries and illnesses."

Carlos receives $444 per week for his injuries. This is less than he would have earned had he not been critically injured. But for Carlos, his losses cannot be measured in hourly wage rates or overtime. For Carlos, his biggest loss is the ability to work hard and provide for his family, as he intended when he went *norte* to the United States and to opportunity.

Grain Handling

O beautiful for spacious skies,
For amber waves of grain,
For purple mountain majesties
Above the fruited plain!
America! America!
God shed his grace on thee
And crown thy good with brotherhood
From sea to shining sea!
"AMERICA THE BEAUTIFUL"

The "amber waves of grain" describe the wheat fields that inspired Katharine Lee Bates to pen her ode as she traveled in the summer of 1893 to

Colorado Springs from the outskirts of Boston, where she taught English at Wellesley College. Anyone who has similarly traveled through America's heartland in the summer months can't help but be in awe of the bounty that Bates saw in farm fields chock-full of wheat, corn, oats, barley, and soybeans. Travelers crossing the "fruited plain" will also see the iconic grain silos, bins, and elevators that sit along railroad tracks and store our plenty on the way to market, or convey the grain to a nearby mill for processing. What is not seen are the workers inside these structures, and the unsafe working conditions that can kill them in a matter of a few minutes.

Grain bins, silos, and elevators are constructed and designed for different storage purposes. A silo is a tall, skinny structure, described as like a coffee thermos that seals tight and holds moisture. Silos traditionally store silage, which is grass or other fodder harvested green and used to feed dairy cattle. Bins, on the other hand, are stubby corrugated steel structures that store dry grain, mostly corn and soybeans. Finally, elevators are the largest storage structures, comprising tall vertical bins, usually concrete, up to 150 feet high and topped by a head house. Inside are a series of vertical conveyors that move grain from the bottom, or boot, up to the head house, where it is conveyed to the concrete bins for storage.

For the workers in grain handling, it is a dangerous and sometimes deadly job, belied by the bucolic and rural setting. Grain dust explosions and fires are a known hazard. OSHA recently reported that over the last thirty-five years, there have been over five hundred explosions in grain-handling facilities, killing more than 180 persons and injuring more than 675.[74] Workers die from falls in grain elevators, suffer amputations from mechanical equipment, and become sick, sometimes seriously, from the gases emitted by decaying or fermenting grains and grasses, as well as from toxic chemicals.

But of all the hazards grain workers are exposed to, suffocation is the leading cause of death. The corn in a grain bin frequently becomes clumped and crusted together from moisture or mold, making it difficult for the grain to flow out through a sump hole at the bottom of the bin. When this occurs, the safe practice is to break up the crusted grain from the outside using a long pole. However, and inexplicably, workers often enter grain bins to break up the corn by walking on the crusted "bridges." When this happens, the "bridge" collapses and the worker is sucked down into the free-flowing grain and completely entombed by corn kernels. The worker will quickly die if not rescued. It is like quicksand, but with corn. The worker will literally drown from the pressure of the compressed kernels on the diaphragm, which makes it impossible to inhale and breathe. This is called compressional asphyxia. If this does not kill, the worker will die from suffocation, much like a person drowning in a body of water: inhaled corn kernels will fill up the nose and mouth, and in two minutes the worker will lose consciousness from a lack of oxygen. In around five minutes, the brain will have depleted all the remaining oxygen in the blood from a still-beating heart, and the worker will suffer irreversible brain damage.[75]

There are well-known safe practices that can save grain-bin workers from suffocating to death and enable their rescue. The safest practice is to provide a worker with a safety harness with a lifeline, the length of which will prevent the worker from sinking any farther than waist-deep in grain. There are still other safety precautions, including using a buddy system, where a coworker is nearby to help if able, and to sound the alarm. Hopefully, in the case of complete entrapment, there is safety equipment nearby, typically a special plastic tube that is forced into the grain and around the worker, while rescuers attempt to extract him.[76]

Still, even with the known dangers and safety practices, grain workers die every year. In 2010, there were fifty-one entrapments, resulting in twenty-six fatalities, a record number. A Purdue University study reported in 2010 that the trend in grain entrapments has continued to increase from year to year.[77] The report also noted that, as with other workplace incidents, the numbers do not reflect all grain-related entrapments, owing to underreporting and the lack of a comprehensive reporting system. The Purdue report estimated that the actual number could be 20 to 30 percent higher. Significantly, 70 percent of all entrapments occur on farms exempt from OSHA's grain-handling standard. Finally, the report found that in 2010, 12 percent of workers entrapped were under the age of sixteen, with the youngest being only seven years old.[78]

The corn crops in recent years have been bountiful. According to the USDA, 2014 was a record year for corn and soybeans, besting 2013's record year. As a result of the record crops, together with the increased demand for corn for food products and ethanol, corn is being stored for longer periods of time. This in turn leads to corn becoming crusted and clumped.[79] Workers breaking crusted "bridges" are the leading cause of grain-bin entrapments. In 2014 there were no fewer than thirty-eight grain bin entrapments.[80] Again, this is likely not the true number, since small farms are outside OSHA's jurisdiction, and self-reporting is unreliable.

In the end, America's "fruited plain" should be a continued reason for celebration. Americans are blessed with a seemingly inexhaustible supply of food. But instead of cause for celebration, the "amber waves of grain" are all too frequently a cause of death to a worker in another preventable incident.

Robert Fitch

In her 2009 letter to shareholders, Archer Daniels Midland chairman, CEO, and president Patricia Woertz wrote, "In fiscal 2009, ADM continued making meaningful progress toward our safety goal of zero incidents, zero injuries," noting a reduction in its "recordable incident rate" from fiscal 2008. This was intended to be good news to ADM's shareholders, along with her report that ADM had extended its record of "uninterrupted dividend payments to 77 years." But for Bob Fitch's family and friends, the news from ADM in fiscal 2009 was not as sunny. For them, ADM's goal of "zero incidents, zero injuries" was hollow corporate spin.

Bob was born in Pensacola, where his father (also Robert) had been stationed while in the navy. Robert and Joyce Fitch had five children. Bob was the middle child. When Bob was fifteen he moved up to Omaha with his father, who was from there and had found work at ADM. Soon after, his sister Cindy and her husband, David Malley, followed them. In August 1977, Bob and Dave both hired on the same day at ADM's South Street milling facility in Lincoln. Bob was nineteen and had worked seasonal construction since leaving high school. Dave was older, at twenty-two. They worked the second shift at the mill, which made feed for cattle, dogs, and horses. Bob eventually became a machine tender, setting up the machines for whatever the plant was making at the time. He did this job for around twenty years. Bob and Dave worked seven days a week, and worked as much overtime as they could get.

When they weren't working, they rode motorcycles and swapped cars, a Corvette for a '65 Chevy pickup. Bob and Dave were close. At their weddings, they were each best man for the other. Bob married in 1982 and had two children, a daughter and son. Dave said that Bob "worked hard and played hard" but was a "good family man" and always found time for his kids, as well as his nieces and nephews. Tonya Ford, Dave and Cindy's daughter, said that for Uncle Bobby, his family was a "package deal."

For around the last ten years, Bob and Dave worked in one of the grain elevators at the South Street milling facility. Grain elevators generally all work the same way. They are usually 70 to 120 feet tall and topped by a head house, where the scale at South Street was located. Inside there are vertical storage spaces with grain bins of various sizes, a work floor, and a receiving pit where trucks unload their grain. Conveyor belts move the grain to other buildings for processing. Workers inside the elevator move from floor to floor using a "manlift," a small platform about the size of a placemat. The manlift runs along vertical rails, only a few feet from the inside grain silo wall. As it moves past each floor in the elevator, the manlift travels through a small hole, known as a manhole. The manlift at the ADM facility was not designed with a safety harness, or a guard railing. Instead, a worker stood on the small platform facing the rails and held on for dear life by a tiny handle.

January 29, 2009, was a frigid morning at the South Street facility. Shortly before 9 a.m., Bob announced over his radio that he was going on a break. He stepped onto a manlift for a ride down the elevator. Dave said Bob most likely put his radio in his pocket because it was cold inside the elevator. But the radio was still connected by a cord to the microphone attached to Bob's collar. Dave thinks that the cord got caught on something and pulled out the radio, hitting Bob in the face and knocking him off the manlift. Bob initially fell more than forty feet, bouncing off the side of the wall. He then smashed into some ductwork. The force of this crash was so great that it disconnected the ductwork from the wall. Bob then slid off the ductwork and down through a manhole, falling forty-four feet more before smashing into the first-floor concrete surface. His radio was found hanging on a bracket attached to the manlift, while his glasses were found on the floor just before his final fall. Bob landed face down, and he lay there in a pool of blood for approximately seven minutes before he was found.

Dave first noticed Bob's glasses on the floor. Then he looked down through the manhole. Dave was the first person to get to Bob. He tried to give him CPR but said "every bone in his body was broken except his little finger," and his "head was laid open." Bob died on impact.

At the hospital, the family was allowed to view the body but could not touch Bob because he was considered evidence in the accident investigation. At the funeral, his body was being held by the funeral home for an autopsy, and there was no casket. The family was allowed to view Bob again at the funeral home but first had to sign a release that they wouldn't sue, including for trauma.

The week earlier, Bob told Cindy that he was afraid of getting hurt at work. Cindy told Dave to protect her brother. Dave has difficulty talking about Bob's death. It was hard for him to go back to work and to ride the manlifts.

OSHA investigated the accident and cited ADM for two "serious" violations, and one "other." The total proposed penalty was $12,500. Instead of paying the penalties and, more important, remediating the hazard that killed Bob Fitch, ADM challenged the citations and penalties. In an informal settlement, OSHA deleted the "serious" citations and penalties. The "other" citation remained, and ADM was penalized $2,500. As for the settlement, OSHA explained that a regulation issued in 1971 specifically identified manlifts as a safety hazard if they did not meet the design requirements of the American National Safety Standard for Manlifts. The manlifts at ADM's South Street facility, including the one Bob was riding, did not meet the safety standards. But the OSHA regulation "grandfathered," or exempted, manlifts installed prior to 1971. Undoubtedly, the grain elevator industry objected to replacing, or retrofitting, existing and unsafe manlifts as too expensive. So the OSHA rule was only prospective, with no "sunset" clause, or date by which the industry was given to bring old manlifts up to current safety standards. Even though the manlifts at South Street were a regulated safety hazard for thirty-eight years, OSHA concluded in Bob's case that the challenged citations would not be upheld because of the exemption. Instead, and as part of the settlement, ADM agreed to replace the manlift that Bob fell from. However, the other unsafe manlifts at the South Street facility, and at other ADM facilities, could still operate.

Following another inspection at South Street, unrelated to Bob's death, OSHA issued a citation on June 1, 2010, that the manlift in elevator D was unsafe because it was not guarded and therefore failed to protect employees. Again, ADM could have voluntarily replaced or retrofitted the old and dangerous manlifts, instead of exposing its workers to unnecessary hazards. But it didn't. ADM challenged this citation and the proposed $7,000 penalty. As OSHA had predicted in Bob's case, bound by the 1971 regulation, an administrative law judge ruled in ADM's favor that the manlift installed in 1954 was not subject to regulation.

ADM's failure to voluntarily replace or retrofit the dangerous manlifts that killed Bob Fitch belied its commitment to employee safety that its chairman professed in 2009 to its shareholders of "zero incidents, zero injuries." For a company with net earnings in 2009 of $1.7 billion, and which rightly boasted that it is "one of the world's largest agricultural processors and food ingredient providers," the failure to protect Bob Fitch and other workers from a preventable accident was inexcusable.

In the end, Bob Fitch's death left his family and friends grief-stricken. Dave Malley simply said, "I miss Bob." For Bob's niece, Tonya, her life changed beyond the loss of Uncle Bobby. Tonya became a worker safety advocate. She testified before Congress in 2010 on OSHA reform legislation. In 2015, Tonya became the deputy director for United Support and Memorial for Workplace Fatalities, an organization that counsels other families who have lost a loved one to another preventable accident. As for ADM, Tonya met with and wrote to OSHA officials, urging them to repeal the grandfather clause. Tonya created a Facebook page for grain mill accidents, which include fatalities from grain bin accidents. In Nebraska, she helped organize public memorials and vigils to Nebraskans who have died in workplace accidents. Her work has not gone unnoticed at ADM. In July 2015, ADM agreed to Tonya's request that it fund the purchase of two grain rescue tubes. The grain rescue tubes will be donated to local fire departments that often lack this important lifesaving equipment when a worker is trapped in a grain bin. The grain rescue tubes were donated by ADM in honor of Bob Fitch.

Registered Nurse

> If a nurse declines to do these kinds of things for her patient,
> "because it is not her business," I should say that nursing was not
> her calling. I have seen surgical "sisters," women whose hands
> were worth to them two or three guineas a-week, down upon their
> knees scouring a room or hut, because they thought it otherwise
> not fit for their patients to go into. I am far from wishing nurses to
> scour. It is a waste of power. But I do say that these women had the

> true nurse-calling—the good of their sick first, and second only the
> consideration what it was their "place" to do—and that women
> who wait for the housemaid to do this, or for the charwoman to do
> that, when their patients are suffering, have not the making
> of a nurse in them.
>
> FLORENCE NIGHTINGALE, *NOTES ON NURSING: WHAT IT IS,*
> *AND WHAT IT IS NOT*

Florence Nightingale, an upper-class English woman, is universally acknowledged as the founder of modern nursing. She trained and supervised
a group of volunteer women during the Crimean War, as well as Catholic
nuns. Her focus on improving the horrid sanitary conditions in field hospitals, and implementing hygiene practices, including hand washing, were
attributed with reducing the death rate among British soldiers. Later, she
helped found the first professional nursing school, in London. Above all,
as she wrote in *Notes on Nursing*,[81] the first responsibility of a nurse is to
care for the patient's suffering, even at the nurse's own expense; and in this
regard, not much has changed.

Today, nurses and other bedside caregivers face a different set of challenges from those of Florence Nightingale and the other women who cared
for wounded soldiers in Crimea. Today, health care workers in hospitals,
nursing homes, and home health care face numerous physical and mental
health issues associated with their jobs. Musculoskeletal disorders prin-
cipally caused by patient handling, and the rate of injury requiring days
away from work, are nearly six times higher than for average workers.[82]
The economic costs of MSDs, especially back injuries, among health care
workers are estimated at around $7 billion annually.[83] In June 2013, U.S.
Representative John Conyers of Michigan and sixteen other members of
Congress reintroduced legislation requiring the secretary of labor to issue
a regulation designed to reduce injuries to health care workers, focusing
on promulgating a safe patient handling, mobility, and injury prevention
standard. Like other OSHA regulations in recent years, this legislation,
previously introduced in 2006 and 2009, died without even a hearing.[84]

MSDs are not the only hazard health care workers routinely experience
at their jobs. In 2014, almost 80 percent of nurses reported being assaulted
while on the job. It is estimated that 70 to 74 percent of all workplace
assaults occur in the health care and social services settings.[85] According
to nurse advocates, more than nineteen thousand health care workers are

assaulted annually, or one every twenty-eight minutes. From 2002 to 2013, the rate of workplace violence in health care facilities was four times greater than in private industry on average.[86] Nurses are hit, kicked, scratched, and spat upon, mostly by patients who may suffer from dementia or are simply frightened and alone and physically lash out at the very people charged with their care and well-being. In scenes reminiscent of the popular TV show *ER*, other instances of assault commonly occur from family members, friends, or even from rival gang members who follow an injured gang member to the emergency room in order to finish him off. OSHA has issued guidelines for violence-prevention programs,[87] but it has not proposed a workplace violence standard, and employers, including hospitals, are left to decide whether to spend the time and money on voluntary programs and training. In early 2014, labor and health professional advocates petitioned Cal/OSHA for a comprehensive work standard on workplace violence in health care facilities. The petition was accepted, and a proposed rule, Workplace Violence Prevention in Health Care, was adopted effective on April 1, 2017, making California the only state with an enforceable standard whose goal is to prevent workplace violence to health care workers.

Work-acquired infectious diseases are among the top hazards all health care workers face. But nurses are exposed at a higher rate, since they are the health care workers in the closest and most prolonged contact with their sick patients. In addition to needle sticks and blood-borne pathogens, occupational exposure to infectious diseases among nurses happens every possible way: by skin and mucous, by inhalation, by eye, and hand-to-mouth. The list of pathogens known to have caused work-related infections is long and growing longer. Victims of outbreaks elsewhere around the globe are transported to the United States and treated by nurses here. Exposure is particularly acute for nurses involved with treating patients with SARS, MERS, and most recently ebola. During the 2014 ebola outbreak, nurses and other health care workers at Texas Health Presbyterian Hospital in Dallas were the canaries in the coal mine when they were treating Thomas Eric Duncan, who became the first person to die in the United States from ebola after he traveled to Dallas from Liberia, where he had been infected. The hospital was ill prepared to treat Duncan, and nurses initially cared for him without proper protections and protocols in place, including Hazmat suits. Two nurses at Texas Health became infected from exposure, and survived.[88]

A two-year survey (2013–2014) on the incidence of needle-stick injuries reported that 5.6 million health care workers are at risk to exposure from blood-borne pathogens, including HIV and hepatitis C. Despite OSHA's enactment of the Needlestick Safety and Prevention Act of 2001, which mandated safety-engineered devices to reduce needle-stick injuries, the survey reported an increase among health care workers in the "sharps injury" rate. Of the 9,494 blood exposures reported, 73.9 percent were due to sharps injuries, while the rest were from mucocutaneous exposures.[89]

Studies have shown that when staffing falls below safe nurse-to-patient ratios, nurses get hurt trying to do too much too quickly, but good patient-outcomes suffer as well. In March 2015, the journal *Critical Care Medicine* reported a study that correlated higher nurse-to-patient ratios with lower patient deaths in intensive care units.[90] Nurse staffing levels are related to occupational infections as well, according to a NIOSH report.[91] Nurse-to-patient ratios correlate with infection rates for nurses and patients alike.

Long hours, unsafe patient loads, MSDs, exposure to infectious diseases, physical assaults, and mental stress make nursing today a dangerous job. The next time a nurse, or other bed care worker, asks how you are doing, be sure to ask back, "How are *you* doing?" You might be surprised at the answer.

Marti Smith

When Marti was in the eighth grade, she took an aptitude test, and nursing came out on top. Her father's response was, "Don't be a handmaiden to a doctor. Be a doctor." Marti eventually ignored his advice, but her journey into nursing wasn't a straight line from the aptitude test. Instead, Marti traveled along a path well worn by many nonconforming boomers. She tried college at Indiana University in Bloomington. There she met her future husband and, after a while, left school, following him to Yuma, Arizona, where he got a job as a reporter. In Yuma, Marti joined the family business and delivered newspapers in the Arizona desert, often to mailboxes along unpaved roads.

While Marti was in Yuma, her mom's throat cancer, which was in remission, recurred with metastasis. She was admitted to the Indiana University Methodist Hospital in Indianapolis for the "dwindles." Marti said,

"She was dying, and went back to the hospital for failure to thrive." Marti returned to Indianapolis to care for her "best friend." At the hospital, Marti befriended a nurse and confided in her that she was planning on getting married. The nurse told her not to wait: "Get married now." So Marti got married in the chapel at the Methodist Hospital. Her mother was there. The advice and compassion from the nurse caring for her mother left a huge impression on Marti.

After her mother died, Marti and her husband moved to Ventura, California, where he found work at a newspaper. They started a family, and eventually two sons were born. For a while, Marti worked at a shelter for abused, mostly working poor, women. Finally, in 1990, she enrolled at Delta College in Stockton to become a nurse. The arc that began in the eighth grade was complete.

Marti became particularly interested in learning about the physiology of the patient, in understanding how the various body parts function and interact. Beyond understanding basic physiology, Marti wanted to know, for instance, that "if a patient's sugar level is 585, what is the potassium?" She didn't want to be a "handmaiden to a doctor," simply filling and following a doctor's orders. She wanted to be a partner in the care and treatment of her patients. If a doctor writes a wrong prescription, it is on the nurse, as well.

After nursing school, Marti became an intensive care nurse. The work was intensely physical and cerebral, requiring knowledge of chemistry and physiology, as well as compassion in dealing with families of critically sick patients. The emotions of the job often turned out to be the most challenging. Marti recalled caring for a nineteen-year-old boy who was in a coma from a horrific car accident while driving drunk and killing another person. The boy's father was so despondent that he told Marti he hoped his son didn't wake up. Marti worked intensive care for nine years at hospitals in California with very high acuity, meaning patients who were very sick. "I liked caring for very sick patients. It was the most challenging," she said.

After intensive care, Marti transferred to the emergency room. There she treated a whole different patient demographic. The ER saw a lot of criminals on their way to jail. But most difficult were the child-abuse cases. The ER is where abused children come. As a result, ER nurses "see some of the worst cases imaginable." Marti recalled a four-year-old girl who was admitted on Halloween. "She was beat so hard, covered with old and new bruises, cigarette burns and cord marks." Marti said the little girl was covered with so many signs of trauma that the only way she could document the abuse was with a rape kit. She explained that the rape kit had a full anatomical diagram that is used to document the places on a rape victim's body that had been attacked. Marti drew all over the diagram, marking the burns, bruises, and cord marks. During her time in the ER, Marti talked about a year where there was a string of child-abuse cases, which included five children who died. She remembered one young child, brought in by the mother, with a fractured femur, or thigh bone. The ER staff suspected abuse, but it was explained away as an accident. Five months later, the same child returned, this time with a fractured humerus, the long bone in the arm. Second "accidents" of this nature are not co-incidence. It is abuse. Because of this, and many other cases, working in the ER took a mental toll on Marti—a work-related injury as acute and cumulative as the MSD injuries she suffered as well. "I felt like I was made of crystal. Like I would shatter," she said.

Marti was never diagnosed or treated for the mental harm that working in the ER caused her, which she described as a "tax on your soul." Nurses and doctors, especially those in the ER, seldom have the time or space to absorb the trauma they have treated—not until later. Marti recalled the time that a two-month-old infant who was in day care died of sudden infant death syndrome. When Marti couldn't resuscitate the baby, she declared the time of death, completed her charting, and went back to work. She had no time to offer condolences, or comfort the family. She had no time to process this tragedy. Instead, she had to tuck away the sadness.

Vicarious trauma, also known as compassion fatigue, is the term used to describe the "cost of caring" for others.[92] It is the trauma of trauma. Other descriptions are "secondary traumatic stress" and "secondary victimiza-tion." In the book *Reflecting on the Concept of Compassion Fatigue*, the author wrote that there is "little doubt that their [nurses'] work may take a toll on their psychosocial and physical health and well-being."[93] Work-place interventions are at the head of the list among the responses to compassion fatigue, particularly for nurses. Yet few facilities or health care systems currently integrate these options into daily operations. Marti's hospital was no exception, and why should it be? There is no recognition at OSHA for compassion fatigue, or vicarious trauma, as an occupational hazard, except in the narrow context of workplace violence. It is largely an unreported and untreated work hazard that has gone ignored, even by nurses themselves. Nurses are focused on the needs of their patients, too often ignoring their own. The drill is just to move on to the next bed.

To be sure, Marti also suffered from multiple MSDs from working as a nurse. She has had injuries from this hazard diagnosed and treated. The physical requirements of nursing are many and challenging. "You can't overstate how physical the work is," Marti said. Her first injury was from a confused patient who yanked on her thumb, pulling it back and tearing her tendon. When nurses went to twelve-hour shifts, Marti developed plantar fasciitis from being on her feet for extended periods. She worked with a stabbing pain at every step.

Consistent with putting patients first, Marti has manually lifted, by herself, many patients who are like a deadweight and unable to assist. In a perfect nursing world, she would have waited for assistance. But in the real world, with staff shortages driven by profit and nurses caring for too many patients at one time, waiting for assistance from another nurse is often not a realistic solution, given the urgent need to move a patient. Similarly, waiting for a mechanical lift is often not an option, as a lift may be in use and unavailable when needed. Or try pushing down the hallway a patient more than two hundred pounds in a two-hundred-pound hospital bed, with a ventilator machine and IV stand attached. Changing bed sheets exposes nurses to back injuries, as well as to fecal matter, blood, vomit, and "confused patients who will fight you like the devil." It is hard work. The very nature of emergent patient needs requires a nurse to act with deliberative haste and not always wait for the safest option. As a result, Marti has had extensive left-shoulder surgery directly related to her many shoulder injuries, as well as knee sprains and a bicep labrum tear, and has severe arthritis in her back. Ibuprofen is her sleeping pill. "I'll know when I'm dead. I won't be hurting anymore," she said. Marti pushed aside her pain and didn't report most of her injuries. "Nurses are caregivers for others. Their physical needs are the last to be treated."

Like many nurses, Marti was exposed to infectious diseases. She said, "I haven't taken antibiotics for ten years. My immune system has been exposed to every disease imaginable," including pertussis and meningitis. With a family at home, Marti would often Hazmat herself after a work shift: she kept a robe on a hook in the laundry room and stripped down and changed on the spot before showering.

Initially, while working at Sutter Hospital in California, Marti tried organizing the nurses there. Staffing was a big issue. There were twenty-six "contract" nurses working in the ER at Sutter. This was not a good, or safe, work environment. Still, the union lost the election. Afterward, Marti moved to another hospital in Modesto. There the union was successful in organizing the nurses, and Marti became the chief nurse steward. In the end, and because of her own experiences, Marti left nursing. She is now a labor representative for National Nurses United, the largest union in the United States for registered nurses. Marti continues to care, this time by representing and advocating for nurses.

Elevators

> So Jack climbed and he climbed and he climbed and he climbed
> and he climbed and he climbed and he climbed till at last he
> reached the sky.
>
> JACK AND THE BEANSTALK

In 2016, Burj Khalifa in Dubai was the world's tallest building at 2,723 feet up to its tip. It has 163 floors aboveground, which are served by fifty-eight elevators.[94] By the close of 2016, more than a half dozen buildings were expected to be completed that top out north of three hundred meters (984 feet).[95] While the sky is literally the limit for these superbuildings, none of these achievements, including the more earthbound buildings framing a modern city skyline, could be possible without the elevator.

Before the modern elevator, there were human- or animal-powered lifts and hoists as far back as the ancient world. Gladiators and animals rode crude elevators up to the arena floor at the Roman Colosseum. Early modern elevators were powered by motors. Still, there remained the ever-present danger of rope or cable failure that would send the hoist or elevator plummeting to the ground along with its passengers and contents. This design flaw was fixed by Elisha Otis in 1852 with the invention of the safety brake, which would engage if the cable snapped. Otis and his new company, the Otis Elevator Company, pioneered all sorts of elevator innovations, and the world grew vertical.

As the modern elevator (and its cousin the escalator) helped engineers and architects push the envelope on building height, incidents involving serious injury and death grew as well. Incidents involving elevators and escalators each year kill around thirty-one people and seriously injure about seventeen thousand people in the United States, according to BLS data. Elevators cause around 90 percent of the deaths and 60 percent of the serious injuries. Half these deaths every year are workers installing, repairing, and maintaining elevators, and working in or near elevator shafts. Of these deaths, 56 percent are the result of falls into the shaft. Other incidents occur when workers are caught in between moving parts of elevators, resulting in 18 percent of deaths, often by decapitation, while workers struck by elevators or counterweights cause 16 percent of deaths.[96] From 1992 to 2009, there were 263 deaths to workers at or near elevators.[97] Of particular note, many fatal falls into elevator shafts occurred when an elevator door opened even though the elevator car was not at a floor. Safety devices preventing a door from opening do not always work, or are deliberately disabled.

These deaths and injuries from elevators are preventable and are not accidents. The lack of routine maintenance and inadequate training for employees are almost always root causes of elevator deaths and injuries. So while we train our gaze upward in amazement at the great buildings that seem poised to pierce the sky, let us not forget the workers who make it possible for all of us to rise safely on these steel and concrete beanstalks as we climb and climb and climb.

John Goldsmith

Ask John Goldsmith what he misses the most about his old life, and he'll say, "I miss being free."

Long Island, New York, stretches over 188 miles into the northern Atlantic Ocean and produces some of the most powerful surf on the East Coast. South-facing beaches are prime locations to ride the Atlantic's swell energy, which consistently results in some of the biggest, best surf on the East Coast.

Growing up in Manhasset, Long Island, John was drawn to the south shore and began to surf the beaches there when he was fifteen years old. He became a "full on" surfer at sixteen. Lido West. Gilgo Beach. And all the way east to Montauk. These were John's playgrounds. John had lots of surfboards. Long, short, and custom boards. And he even won a few dollars from time to time in local surfing contests. He would surf early or late to avoid paying the beach fees, and ate most of his meals on the beach with his friends. John attended St. John's University at the Oakdale campus overlooking the south shore and received a BS in finance. John and his friends would surf before and after classes, and often came to class covered in beach sand. When John was nineteen, he and his surf buddies drove ten hours down to the Outer Banks of North Carolina in search of the perfect wave. They stacked their boards on top of John's pea-green 1968 Plymouth Valiant and headed south. John fondly remembers the misadventures along the way more than the surfing. The Plymouth lost its exhaust an hour out of Long Island. They nearly burned down the place they were staying at by accidentally throwing barbecue ashes into the dumpster. Along the way, John lost his wallet and the car got stuck in the sand and had to be towed out. The experiences John relates seem typical of many college-age kids on their rite-of-passage road trips. But for John they were unique and important, especially now as he scrambles through his memory to recall the great times of his old life.

John also rode motorcycles. He started with dirt bikes and graduated to road bikes after his parents passed away when he was young. He had two Harleys and a Kawasaki "Mean Streak." He rode in all weather, including in the winter with warm riding gloves that his brother T. J. gave him for Christmas. He rode to work and on frequent charity runs like Toys for Tots.

John had an active social life with his friends and family, especially his nieces, whom he spoiled. He was "Uncle Jolly." In addition to riding surfboards and motorcycles, John played hockey and basketball. He was constantly in motion and on the go, until he was not.

After college John worked on Wall Street for many years and later owned mobile phone stores on Long Island. In 2007, he looked to make a change and got a job as the superintendent of a thirteen-floor commercial building at 17–19 Union Square West in Manhattan. John was a member of the Commercial & Residential Building Worker's Union Local 32BJ. He took union-sponsored classes to improve his skills and earned important certificates from the Fire Department of New York in fire-related building systems.

Number 17–19 Union Square West had two manual service elevators. On June 13, 2007, one of these elevators was inoperable and undergoing a complete modernization by Transel Elevator Inc., which had an exclusive contract to inspect and maintain all elevators at 17–19. Building policy required that between noon and 1:00 p.m. the remaining operable elevator would be taken out of service and locked down on the lobby floor while John and the elevator operator took their lunches. No one was to move the elevator car without notifying John during this time. While on his lunch break, John was asked to show potential tenants the penthouse space. Because John could not access the penthouse using the passenger elevator, he took the real estate agent and the prospective tenant to the service lobby where the service elevator was locked down.

In order to open and put the elevator back into service, John needed to use a "drop key," and Transel had provided him one earlier. Normally, a properly maintained elevator parking device will not permit an elevator door to open if the car is not at the floor, even with a drop key. When John and his passengers approached the service elevator door on the lobby floor, he expected the elevator car to be there when he inserted the drop key and stepped in to turn on the light in the back of the car.

John wore his hair on the longer side back then, probably as part of his surfer/cyclist freewheeling persona. So when John opened the door and stepped into the abyss of the elevator shaft, he remembers his hair standing straight up on end, "like in a cartoon," as he plunged around twenty-five feet onto the concrete floor of the subbasement elevator pit. John later described the fall as follows: "When I came down, I landed on my back and my left side, and I hit the ground and heard everything in my body break, all the bones break. I bounced up in the air again and came down and hit again and heard everything break all over again. Then I started screaming." His body was smashed to smithereens. John lay there screaming in pain. Eventually, an ambulance arrived, and he was extracted by firefighters and transported to the Bellevue Hospital Center.

CT scans and X-rays showed that John had suffered severe pelvic and sacral compression fractures. The pelvis is like a girdle of bones that supports the bladder, intestines, and rectum. At Bellevue, and later at Mount Sinai Medical Center, John underwent ten surgeries to try to put him back together, as well as extensive physical therapy.

In the end, John has become a prisoner of his own body, condemned to a life of disfigurement and pain. The inventory of his physical and mental injuries is overwhelming. He walks with a cane, his six-foot, two-inch frame bent forward in pain. He drags his legs behind, especially his left leg, which has no feeling. This causes John to frequently fall when his feet and legs get tangled. He has bowel and bladder urgency, at times uncontrollable, resulting in accidents at least weekly. His body is deformed to where he has described it as, "Looks like I have two heinies." Then there is the constant pain, "wicked, evil," according to John. Walking and standing immediately exacerbate the pain. His most comfortable position is lying on his left side in a fetal position. He has tried every kind of pain medication, all with terrible side effects. At the time of this writing, John takes Dilaudid five times a day to moderate his pain. Dilaudid is a potent Schedule II opioid with adverse effects similar to heroin. Finally, every day John battles depression and PTSD. Most mornings he can hardly muster the strength and will to get out of bed, a process that can take an hour.

Amid all this despair, John found love. He met and married Inna Anderson, and a committed bachelor was no more. They met when John could barely move. Inna works for a major international investment firm. She has three grown kids, and John's stepson now owns one of his surfboards. "She has a heart of gold the size of Montana," John said. Inna takes care of John. She lovingly massages his pain-riddled body.

The New York City Department of Buildings conducted an investigation and concluded that the elevator parking device was "deliberately disabled" by someone who secured the device in the unlocked position by a wire tie. The intentional bypassing of a safety device is a violation of building code. In addition, the investigation revealed that Transel workers rode the service elevator to the twelfth floor to do work, but without telling John or others that the elevator car had been moved from its locked-down position on the lobby floor.

John sued Transel for negligence and eventually settled before trial. Including his lawsuit, in 2011 it was reported that there were eight lawsuits filed against Transel.[98] On December 14, 2011, Suzanne Hart, an advertising executive, was killed in a Transel-maintained elevator after she entered an elevator and the door suddenly closed on her, dragging her body up. John is prohibited by a confidentiality agreement from discussing the terms of the settlement. However, this much is known: there is no amount of money that can compensate John Goldsmith for the life that was taken from him. "I was doing everything," he said. Barbecues, parties, surfing, motorcycles, friends, and his family. Now, when Inna is at work, he sits at home and watches a lot of court TV. Occasionally, a retired friend stops by for a little company. To just get up and go to see friends or family, something he never gave any thought to before, is now dependent entirely on how his body feels that day. Any unwillingness to get up and go is not because John's friends have stopped loving him, which they haven't. But he doesn't want to be an anchor to his friends. Life goes on, sometimes without him.

"It all stopped instantly," John said. "Everything just stopped."

Postscript: I finished writing John's story in September 2016. I promised John that I would get out to Long Island that fall to meet him in person, because his was the only interview that I did by phone, over the course of several conversations, as well as by e-mail. However, my day job got in the way, along with other excuses, and I didn't make the visit, although I believed I would have plenty of time to keep my promise. In late January 2017 I learned that John had passed away. John had been struggling with kidney issues for some time, as a result of his injuries. His sister Karen reported to me that in early January John went into the hospital because of his kidney problems and never came out. John was moved to the ICU, intubated and put on a respirator. Eventually he was removed from the respirator, and passed on January 24 from multiple organ failure. There was a service for John, which Karen reported was packed with family and friends from all stages of his life. His friends from elementary school, surfing, and biking were there to say goodbye. Karen said the service was filled with a lot of love for John. While I missed my chance to see John, I know from Karen and from John how pleased he was with his story. He finally got to talk about how his life was unalterably changed forever. He got things off his chest. In my conversations with John, he was angry and frustrated at times, but was mostly upbeat under the circumstances, and never asked for sympathy or to be cast as a victim. I am honored to tell John's story.

6

WHAT CAN WE DO?

Throughout the writing of this book, I have struggled with the challenge posed at the beginning: that we must do everything possible to make Americans safe at their workplaces. But what is the measure of that challenge? How can we know when we have done everything possible? Whereas there is no doubt that reported deaths and injuries have declined over the years, is there a number that is acceptable as the cost of doing business? At what point can we say that we have done enough? Is one preventable death acceptable? Are more rules and regulations going to move the statistical needle? Or, without meaningful deterrence and enforcement, are more rules and regulations just thousands of words on the pages of the *Federal Register*? And does protecting American workers extend beyond their physical safety at a work site, to protecting and caring for the workers, and their families, when they become injured, disabled, or die?

Before addressing these questions, it is important to consider the arguments of those who oppose robust and effective health and safety laws and a larger role for government oversight and leadership. To begin with, there is a popular, if not prevailing, view in America that throws up its hands and reduces the complexity of worker safety to the simple notion that life is inherently and randomly dangerous, and we cannot be expected to protect every worker, nor should we. Accidents happen. Trains derail. Cars crash. Workers get injured and die at their jobs.

This fatalism has been widely appropriated by politicians and policy makers as the go-to rebuttal to practically any social problem that could be fixed or ameliorated by the hand of government. This do-nothing view is most often expressed in the gun debate; any type of gun control regulation will not stop all killings, so why propose new regulations at all? It is the marching beat of the big-government-destroys-our-freedoms crowd. Workers and worker safety have not escaped the broad brush of this public policy nihilism. But it is an inherently unsound and immoral argument, especially when it comes to worker health and safety.

Two prominent conservative think tanks, the Cato Institute and the Mercatus Center at George Mason University, are archetypical of those who espouse and promote this "antigovernment organizing" philosophy and in turn lobby politicians to codify its principles.[1] The Cato Institute was founded by the Charles Koch Foundation.[2] David Koch is a member of Cato's board of directors.[3] When it comes to worker safety, Cato has argued that free markets "have done much better than governments at providing safety" for workers.[4] Founded in 1980, the Mercatus Center describes itself as a university-based research center for "market-oriented ideas."[5] Mercatus board of directors member Richard Fink is also an executive vice president of Koch Industries Inc. and a member of its board of directors.[6] In 1978, he founded Mercatus, which was briefly housed at Rutgers University before relocating to George Mason in 1980.[7] Charles Koch also sits on Mercatus's board of directors.[8] Since 1985, Mercatus and George Mason have received more than $30 million from Koch foundations.[9]

Cato publishes a quarterly journal, *Regulation*, which boasts that it examines nearly every market "and nearly every government regulation."[10] Its contributors and editorial board claim the most prominent conservative thinkers and academicians in America, including the late Supreme Court associate justice Antonin Scalia. In 1995, *Regulation* published an article with the make-no-mistake-about-it title "Abolishing OSHA." The authors argued that "OSHA can never be expected to be effective in promoting worker safety; that an expanded OSHA will cost jobs as well as taxpayer dollars; and that other means currently keep workplace deaths and injuries low and can reduce them even more."[11] This conclusion neatly fits into Cato's antigovernment and antiregulatory construct, but it is badly flawed.

One of the most inconvenient truths standing in the way of this argument is the indisputable fact that since OSHA's inception in 1970, workplace deaths, as of 2013, have decreased more than 66 percent. During this same period, occupational injuries and illnesses have declined by 67 percent. All this occurred at a time when the United States workforce nearly doubled in size. Fatalities went from thirty-eight workers per day in 1970 to thirteen per day in 2014. As of 2014, injuries and illness are down to 3.3 per 100 workers from 11 per 100 in 1970.[12] How does "Abolishing OSHA" coauthor John Leeth deal with these facts? He argues that the decline during this time period has more to do with temporal coincidence than anything attributable to OSHA. And, faced with more than forty years of data, Leeth asserts in a 2013 Mercatus article that OSHA is "not the major cause" of the decline. However, he did begrudgingly admit that OSHA has had a "modest" role in improving worker safety.[13]

This would be a major concession, albeit belated, from the coauthor of "Abolishing OSHA," except that the "modest" role might have been even greater if OSHA had been given more resources to protect workers, and not less. In a revealing moment of tone deafness, typical of those who spin facts to fit their belief systems, and without considering an alternative reality, Leeth writes that OSHA's inspection efforts have reduced injuries by only 4 percent.[14] Stop and take a moment to think about that statement, made in support of the ineffectiveness of OSHA. Then consider the fact that OSHA is armed with very few inspectors: there are only about twenty-two hundred inspectors responsible for the health and safety of 130 million workers, employed at more than eight million worksites around the nation—which translates to around one compliance officer for every fifty-nine thousand workers.[15] Given this minuscule enforcement resource, it is a wonder that inspections have reduced injuries by even as little as 4 percent.[16] Rather than argue for more inspections, the do-nothing "antigovernmenters" want fewer inspections, even as the data indisputably show that employers who have been inspected have fewer safety problems later on. A study of more than half a million OSHA inspections found total violations decreased by 28–48 percent from the initial OSHA inspection to the second.[17]

Even for Cato and Mercatus acolytes it is tough to continue to insist that abolishing OSHA is a good thing, and that "it can never be expected to be effective in promoting worker safety," when faced with hard facts that assert otherwise.[18] Thus, the updated view offered by Cato, Mercatus, and others is that "OSHA can best complement the other pillars of the US policy system by providing information to workers about possible hazards, particularly health-related hazards, and by gearing inspections toward worksites where dangers are hard to monitor and firms employing less mobile and less knowledgeable workers."[19] This hard-won acknowledgment is long overdue, but is tempered by the persistent denigration of OSHA's funding and rule-making authority. Instead, it is argued that OSHA is merely a complement to the other pillars of worker safety policy.[20] It is important to note that no worker safety advocates have ever argued that OSHA is the only pillar in worker safety policy. Rather, they assert that OSHA is part of a comprehensive occupational health and safety regime.[21] Nonetheless, let us examine a policy system that the other pillars of the market-oriented proponents claim will protect American workers.

A 2008 Cato policy report authored by David Henderson argues that "free markets have done much better than government at providing [worker] safety, fairness, economic security, and environmental sustainability."[22] Simply stated, there is nothing that this single-source solution cannot fix if government would just get out of its way. But in regard to workplace safety, Henderson writes, "in short, there is and has been a 'market for safety.'"[23] Leeth similarly argues that the labor market is one of the four pillars in the U.S. worker safety policy system: "The positive relationship between wages and risk means firms with better safety records are rewarded by the market by being able to pay less to attract equally qualified workers than firms with worse safety records."[24] Let us see if this statement holds true.

The argument is that workers demand job safety "by the wage premium we insist on to take a given risk." As the so-called risk wage premium rose, employers found it cheaper to avoid the risk premium by increasing safety in the workplace.[25] One of the many fallacies of this argument is the assumption that workers are aware of the risks in their workplace, and that armed with this information, they can make an informed, rational decision about whether to work there based on an employer's safety record, and at what price. This is an enormous assumption that, in the real world, simply does not hold true. We know that employers and employees underreport in substantial numbers. Accordingly, true and accurate information is simply unavailable.[26] Moreover, in almost all instances, workers simply do not possess the information to demand a wage premium based on an assessed risk, accurate or otherwise. Even if an employer's real safety record is known to a worker when completing a job application while sitting in the human resources office, workers are not insurance companies—they are unable to underwrite their expected wage rate based on the safety risk of their prospective employer. They simply are not armed with the ability to handicap their job risk, nor do they possess the bargaining power to leverage and "insist" on anything with a prospective employer.[27] As will be discussed, the inequality of bargaining power is even greater for nonunion workers.

Regardless, by the very nature of a particular job or industry, some risks are patently obvious. Working in an underground coal mine, hundreds of feet into a mountain, poses an obvious risk for which a worker could demand a higher wage. But even then, the most important information is often unknown. For example, the miners working at the Upper Big Branch mine when it exploded from coal dust combustion may have been unaware of the extent to which their employer was violating mine safety laws, including those designed to prevent mine explosions. More significantly, coal companies have been able to avoid being designated as a "pattern" violator for multiple serious safety violations by dragging out the lengthy citation appeals process.[28] Had the Upper Big Branch miners known the real risks involved with their jobs, there is certainly no wage premium large enough that a reasonable worker would have accepted.

Unless, of course, there are no other jobs tucked into the hollers of West Virginia, or eastern Kentucky, which raises yet another fallacy in the market-oriented argument. In a scenario where demand for jobs outstrips supply, particularly in time periods or industry sectors with high unemployment, basic economics dictate that workers can demand nothing in the way of a wage-risk premium. They take the jobs that are there and at the wage offered. Thus, notwithstanding the inherent dangers in this kind of underground mining, there are not many other jobs available for these workers besides mining or the military.

Moreover, the ability of workers to demand higher wages, whether in return for risk or skill, has historically depended in large part on their power to bargain. In 2012, the *Wall Street Journal* acknowledged the existence of a union wage premium of around 23 percent between workers in right-to-work states versus non-right-to-work states.[29] In union settings, workers can demand a higher wage through their ability to bargain collectively. Not so much for nonunion workers, including those working in the now largely "union free" underground coal mining industry, including the entire state of Kentucky.

Furthermore, the benefits of unionization and collective bargaining extend to safety. Union work sites are safer than nonunion work sites.[30] However, this is not attributable to the union wage premium, but to the negotiated mechanisms and legal protections in place at a union work site. For example, unions often negotiate safety rules and have safety committees. Moreover, workers at a union plant are less fearful to complain about unsafe working conditions than at a nonunion plant. Nowhere is this more evident than in the coal industry, where nonunion miners fear for their jobs if they raise health or safety concerns. The Upper Big Branch was a nonunion mine.[31] According to the United Mine Workers of America, between 2002 and 2010 only around 11 percent of U.S. coal mining fatalities—or 30 fatalities—have occurred at unionized mines.[32]

Finally, the laissez-faire risk-wage premium that is supposed to invisibly protect workers is of little protection to the millions of low-wage workers, many of whom are women of color and undocumented. The risk-wage premium for these workers is virtually nonexistent, particularly for undocumented workers in dangerous industries like construction and landscaping. The threat of deportation and of other penalties for immigration violations naturally reins in a worker's demand for a higher wage in exchange for a safety risk. Additionally, the risk-wage premium offers no protection for the millions of temporary workers who have many "new" jobs each year in different industries and are given little or no safety training at their temporary job. These workers, typically without a union, can be "returned" to the employment agency for a less difficult worker.[33]

In the end, today there is no likelihood that a wage premium can protect American workers adequately, if at all. Instead, in the United States, the workforce is more part time than ever before;[34] and the new temporary worker is waiting to take the place of a traditional, non-contingent worker. Unionization is at 11 percent of the workforce, a historical low-water mark.[35] Against this backdrop of a fractured and vulnerable workforce, the politics and philosophy of laissez-faire fatalism promoted by the ascended plutocracy, championed and funded by the likes of the Koch brothers, has continued to put American workers at great risk, but with no risk-wage premium.

The other two pillars protecting workers, according to Leeth, are the legal and workers' compensation systems.[36] For the free-marketers, these systems work as intended—a true statement—but not necessarily for the good of worker safety.

The legal system, which the free-marketers argue incentivizes employers to provide safe working conditions for fear of financial liability, actually does nothing of the sort. As many of the stories in this book reveal, and as is replicated millions of times for other American workers, the courts are all but closed off to injured workers for claims against their employers by the workers' compensation bar. The workers' compensation system was precisely designed to insulate employers from liability imposed by a jury for death or injuries to their workers. Injured workers, and surviving family members, have no legal recourse against employers, outside of compensation benefits. As such, legal claims can be brought only against third parties, for example the manufacturer of a machine that injured an employee. But these claims are exceedingly difficult and expensive to prove, and are often subject to liability caps that were enacted by politicians hostile to worker safety, and beholden to the likes of the Kochs and their allies.

The final pillar, workers' compensation, is a varied system of state laws that depend on the same anti-worker politicians for their continued existence and benefit levels. Too many workers live and work in states with workers' compensation systems that make it difficult for them to obtain adequate benefits. The dominance of corporate money in politics has resulted in the significant reduction of workers' compensation benefits and the creation of more legal hurdles for workers to clear before they receive these parsimonious benefits.[37] These same politicians tout their failed—for workers—workers' compensation systems as a reason for businesses to relocate to their employer-friendly states. For example, the Indiana Economic Development Corporation, the state's economic development agency, proudly brags on its website about the usual list of pro-business incentives for businesses considering whether to move to Indiana. These include the state's low corporate tax rate, its right-to-work law, its "regulatory freedom," and the fact that "Indiana boasts the second lowest workers' compensation rates in the country."[38] These rates, and lower workers' compensation costs, are attributable in large part to Indiana's lower average weekly wages, thirty-fifth in the nation.[39] Yet despite the low wages, and low workers' compensation rates, in 2014 Indiana had 130 worker fatalities, surpassed only by states with much larger workforces.[40] In the end, this race to the bottom exposes more and more workers to the vagaries of a deeply flawed workers' compensation system in need of a national overhaul, an unlikely outcome in this era of corporate money poisoning our political processes. As such, rather than creating an economic incentive for employers to provide a safe workplace, the workers' compensation system, with its legions of insurance adjusters and their chosen doctors, serves instead to protect and insulate employers from any legal liability and meaningful financial exposure.

Grim as this all sounds, are there any solutions? Can we do better? Yes, and yes. And it is important to keep in mind that there are lots of good ideas from worker safety advocates that can make American workers safer, just not from Cato and Mercatus. No matter how dispiriting it is to be constantly pushing the boulder up the hill, especially against such a well-funded and resourced opposition, if we stop putting forth ideas, then the workers will have lost. That means that we, and our families and friends, will suffer. So, and in no particular order, here are some ideas that might help keep American workers safer.

OSH Act Reform: In 2017, it will be forty-seven years since the Occupational Safety and Health Act became law. In that time, the American workforce has undergone enormous changes that Congress could not have anticipated. There are nearly twice as many workers now than there were in 1970. In 1970, there were 78.5 million Americans working. In 2012, that number was reported by BLS at 143 million.[41] There are more contingent workers than there were. In 2016, the American Staffing Association reported that more than three million temporary and contract workers were "employed" every week by staffing agencies, and annually more than sixteen million temporary and contract workers flow in and out of staffing agencies to America's work sites.[42] While this provides employers with greater flexibility, and arguably shields employers from their

legal obligations to these workers, this revolving door has disastrous consequences for the health and safety of these workers. A Washington State
study showed that temporary workers in the construction and manufacturing industries have twice the rate of injuries of employees in traditional
employment relationships.[43] This is because temporary workers lack adequate safety training; employers are less likely to invest in their safety
training, given that they will be working at a job for only a short time. As
a result, temporary workers are more likely to be injured on the job during
the first weeks of their assignment. They are also assigned the most dangerous jobs, because employers have a financial incentive to contract out
these dangerous jobs to workers "employed" by a staffing agency in order
to avoid higher workers' compensation premiums based on the experience
rating of their employees.[44] The combination of inadequate training and
the assignment to dangerous jobs is a catastrophic formula for workers.
Moreover, the misclassification of employees as independent contractors,
a trend that has reached epidemic proportions, puts workers at risk by
creating economic disincentives for employers to invest in and provide a
safe workplace. They're not my workers, and their safety is not my problem, goes the refrain. Misclassification again allows employers to avoid
higher workers' compensation premiums based on the experience rating
of employees, not to mention the avoidance of paying their share of payroll taxes, Social Security, Medicare, and unemployment withholdings.
The rise in a contingent workforce, and the rampant misclassification of
workers, have unmistakably altered the traditional employer-employee
relationship, creating a "fissured workforce" that in turn has "blur[red]
responsibility for safety and put workers in harm's way."[45] Finally, the nature of work has changed dramatically in the almost half-century since the
OSH Act was passed. America's manufacturing sector has shrunk and has
been replaced by millions of new workers and jobs in the service industry.

Because of these and other changes, the OSH Act needs to be reformed
and updated to make it relevant to this new workforce and economy. First
introduced by Senator Edward Kennedy on Workers' Memorial Day on
April 29, 2004, the Protecting America's Workers Act (PAWA) sought to
enact badly needed reforms to the OSH Act. Senator Kennedy declared,
"On Workers' Memorial Day, we remember those who have died or been
injured on the job in the past year. We renew our commitment to them and
their families to do all we can to end the unsafe and unhealthy conditions

that still plague so many workplaces across America." However, with Congress and the people preoccupied by a disastrous war in Iraq, the Senate bill, and its counterpart in the House, died without a vote. Since then, Democrats in Congress have reintroduced PAWA every Workers' Memorial Day, and still there has not been a vote. With the Republicans in control and in the majority, the bill has not been the subject of any congressional hearing since 2010. The 2015 PAWA bill, reintroduced by Senator Al Franken, would, among other things, expand OSH Act coverage to millions of public-sector workers, increase penalties, including the availability of felony charges against willful and repeat violators, increase protections for whistleblowers who report OSH Act violations, and expand the General Duty Clause to require employers to provide a safe workplace to all workers, including contingent workers, at their work sites.[46]

"Political space" is a common buzzword among politicians. Translated, it means the political cover to do the right thing. Political space usually occurs only when huge catastrophes move public opinion to outrage and threaten the careers of our elected representatives. Think of the 2008 financial meltdown, which provided political space at the time for the passage of Dodd-Frank. On the other hand, if the murder of twenty children at the Sandy Hook Elementary School in Newtown, Connecticut, could not provide the political space for even the most symbolic gun reform, what chance does OSH Act reform have? The website govtrack.us, which tracks bills in Congress, has given the 2015 version of PAWA a 1 percent chance of passage in the House and a slightly better 4 percent chance in the Senate.[47]

Enhanced Civil Penalties and Criminal Prosecutions: There is no single silver bullet for making workers safer. But criminal prosecutions and enhanced penalties may be the closest thing to one.

At a Health, Education, Labor and Pensions (HELP) Senate hearing on April 29, 2008, "Focusing on Penalties Related to Workplace Safety," then-chairman Kennedy began by noting that the median civil penalty for a workplace fatality in 2007 was $3,675, a number that had remained largely unchanged. He then stated, "Workers' lives are obviously worth more than that. Employers who ignore their employees' safety should pay a penalty that will force them to change their negligent ways. It is the only realistic way to save lives. A mild slap on the wrist is not enough." Senator Kennedy went on to compare the civil penalty provisions in other regulatory laws. Violating the South Pacific Tuna Act of 1988 can net a fine of $325,000. A Clean Air Act violation can result in a fine of $270,000, while a violation of the Fluid Milk Promotion Act can cost $130,000. Kennedy added, "Protecting tuna fisheries is important, but so is safeguarding workers' lives and we need to raise OSHA's penalties if we hope to deter unsafe working conditions."[48] In March 2010, Dr. David Michaels, the assistant secretary of labor for occupational safety and health, testified before Congress on the need for enhanced civil penalties. He noted that when a tank full of sulfuric acid exploded at a refinery in 2001, literally dissolving a worker there, the employer was fined only $175,000 by OSHA. But in the same incident, the Environmental Protection Agency fined the employer $10 million for violating the Clean Water Act, when thousands of fish and crabs were discovered dead from the explosion runoff.[49] It is abundantly clear that by applying a free-market analysis to the issue of penalties, there is simply no financial incentive for employers to provide a safe workplace.

There is no genuine dispute that the civil penalty regime in the OSH Act is woefully inadequate as a meaningful deterrent to prevent safety violations. For a "willful violation," Section 17(a) of the OSH Act provides that the maximum civil penalty is $70,000, and the minimum $5,000.[50] Keep in mind that under OSHA's Field Operations Manual, "willful" is the most demanding standard and, for a violation, requires that the employer has demonstrated an intentional disregard for the requirements of the OSH Act or a plain indifference to employee safety and health. The employer either knows that what he or she is doing constitutes a violation, or is aware that a hazardous condition existed and still made no reasonable effort to eliminate it.[51] According to the United States Attorneys' Manual, a "willful" violation is the failure to comply with a safety standard under the OSH Act if done knowingly and purposely by an employer who, having a free will or choice, either intentionally disregards the standard or is plainly indifferent to its requirement. An omission or failure to act is willfully done if done voluntarily and intentionally. Some courts have even permitted an employer to argue ignorance of the law in considering an employer's intent.[52]

For a "serious" violation, a lesser penalty than willful, an employer can be fined up to $7,000. A violation is "serious" where there is substantial probability that death or serious physical harm could result and where the employer knew, or should have known, of the hazard.[53]

Section 17(e) provides that for a willful violation resulting in the death of an employee, an employer "shall, upon conviction" be punished by no more than $10,000, or not more than six months in prison. For a recidivist employer, who is convicted "after a first conviction" for a violation of causing the death of an employee, the OSH Act merely doubles the maximum penalties to $20,000, or one year in jail.[54] Arguably, general criminal penalties in 18 U.S.C. Sec. 3571(c)(4) of the United States Code can apply, with maximum penalties of $250,000 for an individual, or $500,000 for corporations, who willfully cause the death of an individual.[55] However, as will be discussed momentarily, most U.S. Attorneys have been reluctant to prosecute worker death cases, preferring instead to invest their resources in bigger headlines, such as drugs, political corruption, and terrorism cases.

Section 17 has other penalties, not directly related to the death or injury of a worker. If anyone "gives advance notice of any inspection" and tips off an employer so it can cover up a possible workplace hazard, the maximum penalties, for what amounts to an obstruction of justice, are $2,000, or six months in jail. Finally, making false statements, or perjury, in documents that need to be reported under the act, such as injury reports, is punishable by a maximum penalty of $10,000, or six months in jail.[56]

Historically, the maximum fines described above have rarely been applied. In 2007, the average penalty for a serious violation, where there is a substantial probability that death or serious physical harm could occur, was $906. For violations that are "other than serious," the average OSHA fine was only $40. In death cases, the average OSHA penalty assessed in 2007 was just $2,343. The already low maximum penalties, particularly for the most common serious violation, are routinely adjusted downward based on an employer's size and history, often resulting in a reduction of 30 to 70 percent. When challenged, these reduced penalties are often abated further by another 30 to 50 percent.[57]

One study found that the systematic discounting of fines by OSHA in settlement negotiations with an employer threatens to negate the deterrent value of citations. The study found, for example, that during the Obama administration, OSHA settled citation cases at a median of 25 percent below the initial citation in worker fatality cases. In fatality cases, this resulted in median penalties of only $5,800, "less than the cost of an average funeral."[58]

In an all-too-familiar example of the ineffectiveness of OSHA's discounting policy, in June 2016 Sunfield Inc., a motor vehicle metal parts stamping plant, was fined $3.42 million by OSHA for willfully exposing temporary workers to machine hazards. Sunfield is a serial offender, having been inspected twenty times since 1997, resulting in 118 citations.[59] Each time, Sunfield promised to address the unsafe working conditions. Notwithstanding Sunfield's recidivist behavior, it still retains the right to contest the latest fines. For scofflaws, OSHA should deny penalty reductions and requests to reclassify or withdraw citations and send the message that repeatedly gambling with workers' health and safety must come to an end.[60]

With a change in administration at OSHA after the 2008 election, the enforcement emphasis had begun to change. In June 2010, OSHA adopted the Severe Violator Enforcement Program (SVEP), intended to focus enforcement efforts on significant hazards and violations, such as fall dangers, amputation dangers, combustible dust, silica exposure, excavation/trenching dangers, and shipbuilding hazards.[61] In addition, OSHA revised its penalty policies and the factors used to adjust penalties. However, for employers with between one and twenty-five employees, the reduction factor is as high as 60 percent, even though small businesses have high injury rates.[62] Even with a new emphasis, total OSHA work-site inspections remained flat over a five-year period from 2009 to 2013, at around thirty-nine thousand.[63] During the same period, total violations actually fell almost by ten thousand, including in all categories except repeat violations.[64] OSHA claims that the number of violations has declined because of its enhanced and proactive enforcement activity, although the minimum proposed penalty for a serious violation increased only to $500 before abatements, hardly a financial incentive for an employer to provide a safe workplace.[65]

The Section 17 statutory penalties, meager as they are, had been adjusted upward only one time—in 1990—since the OSH Act was passed. Moreover, unlike other federal enforcement agencies, the penalties in the OSH Act have been exempt from the Federal Civil Penalties Inflation Adjustment Act. Not adjusting the penalties for inflation reduced their real dollar value by around 39 percent.[66] The PAWA seeks to correct this by increasing the penalties for serious and willful violations.[67] This increase would only recalibrate the real-dollar value to 1990 values.[68] Nonetheless, there must be a price point for workplace hazards that will incentivize employers to create and operate safe workplaces.

In a sliver of good news, in the Bipartisan Budget Act of 2015 Congress approved an increase in fines that OSHA and other agencies could levy pegged to the Consumer Price Index, and automatically adjusted for inflation going forward. What this means is that civil penalties for serious OSH Act violations could increase to $12,471, from the current maximum amount of $7,000. Willful violations will increase to $124,709, from the current limit of $70,000. Finally, the legislation provides that an agency head can reduce the penalty from the maximum amount based on certain factors, including the all-consuming loophole that the penalty would have a negative economic impact.[69] What penalty doesn't have a negative impact? By definition, any penalty naturally has some negative economic impact. However, the real question is whether these, or any penalties, will help reduce workplace injuries. While it is too soon to know the effect that these long-overdue increases will have on protecting workers, a study in 2015 by the Institute for Work and Health concluded that there is a strong correlation between penalties and citations and a reduced rate of workplace injuries.[70] But, at the very least, the increases may represent a small recognition by Congress that the penalty regime in federal regulatory statutes, particularly the OSH Act, is woefully inadequate as a deterrent to violators.

How about a criminal prosecution? It would be difficult to argue that the threat of meaningful jail time is not a deterrent, except to the sociopath. Yet, as already discussed, under the OSH Act, the maximum jail time for an employer that willfully violates workplace safety laws, and where a worker dies, is only six months. Criminal deterrence should be measured in years, not months. Again, Senator Kennedy: "If you improperly import an exotic bird, you can go to jail for two years. If you deal in counterfeit money, you're looking at 20 years."[71] His point was obvious. Without a credible threat of jail, "many companies treat safety violations as another cost of doing business."[72] The criminal provisions of the OSH Act must be strengthened to protect American workers.

Currently, Section 17(e), 29 U.S.C. Sec. 666(e), provides that a Class B misdemeanor prosecution for an OSH Act violation will occur only if there is (1) a willful violation of worker safety regulations that (2) results in the death of a worker. The closest comparison to Section 17(e) in state and federal criminal codes is involuntary manslaughter, a felony that has a wide range of sentencing alternatives, but none as low as six months for causing the death of a worker.

Beyond the insignificant civil penalties and jail time, the statute as currently written is a failure on many levels. First, using a "willful" standard allows an employer to use its ignorance of the law as a defense to the element of intent, which is contrary to well-established American jurisprudence that ignorance of the law is no defense. Second, changing the standard to "knowing" violations is more in line with other regulatory crimes, such as violations of environmental laws. Under this standard, a defendant need only know of facts or circumstances that are unsafe, and need not know that the unsafe conditions are a violation of a particular safety regulation or law. This also eliminates the ignorance-of-the law defense. Third, permitting prosecutions only where a worker has died ignores the seriousness of most other safety violations, which can result in horrific injuries and lifetime pain and suffering. Moreover, violations that endanger workers, even if no injury or death has occurred, should be criminalized in appropriate circumstances. This is a commonsense and proactive approach that does not wait until a worker gets injured or dies. A criminal worker-endangerment provision would also have a strong deterrent effect where an employer is aware of unsafe work conditions that have not yet injured or killed a worker. Fourth, classifying worker safety crimes involving the death of a worker as only a misdemeanor all but guarantees that prosecutors will not waste their limited resources except for crimes that Congress has designated as felonies. Finally, criminal prosecutions have to be expanded to include supervisors, and not just the "employer." Supervisors are the persons who direct the employees and can place them in harm's way. Putting supervisors at risk for criminal prosecutions might have the added benefit of creating an important layer of whistleblowers with direct knowledge of an employer's internal polices and decisions that create the unsafe workplace in the first place.

Given this list of inherent structural flaws in the OSH Act's criminal regime, it is clear that OSHA has been severely constrained in its ability to keep workers safe through the threat of criminal prosecutions. Between 1970 and 2008, there were 340,000 workplace fatalities. Yet during this same period, there were only sixty-eight prosecutions, resulting in a total of forty-two months in jail.[73] From 2010 through 2013, there were only forty referrals for criminal prosecution, with only three referrals in 2013.[74] And most referrals are declined for prosecution. During this same time, 18,596 workers were killed on the job.[75]

In the end, prosecuting regulatory crimes where workers die has not been a priority for any Department of Justice or U.S. Attorney's office. Worker death cases are simply not headline grabbers that can help catapult an ambitious U.S. Attorney to political fame and glory, or to riches in private practice as a defender of prominent white-collar criminals. Nonetheless, in December 2015, the U.S. Departments of Labor and Justice entered into a memorandum of understanding (MOU) "to provide for coordination of matters pertaining to worker safety that could lead to criminal prosecutions by DOJ."[76] On its face, the MOU looks like a meaningful change in prosecutorial emphasis. But it did not change any statutory enforcement provisions, commit to prosecute any cases, or add any new resources. Instead, it promised merely to share information, to cooperate and coordinate between the departments. Indeed, it clearly states that "nothing in this MOU commits DOJ to investigate or prosecute any particular worker-safety incident."[77]

Other regulatory statutes have stronger criminal provisions. Environmental laws, for example, have meaningful criminal penalties, and nearly every federal and state prosecutor's office has an environmental specialist tasked with investigating and prosecuting environmental crimes. The Clean Water Act (CWA) makes it a crime to negligently discharge a pollutant into a water system of the United States without a permit, or in violation of a permit, punishable by one year in prison.[78] Knowing violations are punishable by three years in jail and fines of up to $100,000 per day.[79] The CWA has many other provisions that criminalize conduct, including a "knowing endangerment" crime that has been used to protect workers from their employer's crimes to the environment. Knowing endangerment makes it a crime if an individual or a corporation "knew at the time that such acts put another in imminent danger of death or serious bodily injury." Conviction of this crime under the CWA carries with it a fifteen-year prison sentence and a $1 million fine for a corporation.[80] While not intended exclusively as worker-safety measures, these "knowing endangerment" crimes under environmental laws, such as the CWA and the Resource Conservation and Recovery Act of 1976, have been used by some federal prosecutors to protect workers from life-threatening environmental conditions created by their employers, something the OSH Act is unable to do.[81]

A legitimate argument is made that these criminal laws are not being used enough as a deterrent against environmental crimes, and that DOJ resources focus more on fighting terrorism and drugs. In a battle of numbers, the Environmental Protection Agency reported that in fiscal year 2013 it assessed more than $1.5 billion in criminal fines and restitution and charged 281 defendants, resulting in 161 years of incarceration.[82] Critics claimed these victories came from a relatively few big-name prosecutions.[83] Nonetheless, the obvious point is that the OSH Act does not even have meaningful criminal laws on the books in the event that enforcement priorities begin to recalibrate. More galling is that there is not even a "knowing endangerment" crime in America's worker safety statute. Protection for workers in that area must come from environmental laws, and increasingly from state prosecutions. Prosecuting worker safety crimes under other statutes, rare as they are, merely underscores the toothlessness of the OSH Act's enforcement provisions. Nonetheless, there are a few notable prosecutions that may send a strong message of deterrence.

In April 2015, Los Angeles prosecutors charged Bumble Bee Foods and two of its plant managers in the death of a sixty-two-year-old worker who was cooked inside a massive oven along with tons of tuna. In August 2015, Bumble Bee settled the criminal charges for $6 million, making it the largest settlement in California for a workplace safety prosecution. As part of the settlement, Bumble Bee agreed to plead guilty to a single criminal count, pay the worker's family $1.5 million, and implement safety measures. The plant managers individually agreed to plead guilty to misdemeanor charges and pay thousands of dollars in fines.[84] Before Bumble Bee, in 2013, there were 189 worker fatality cases criminally investigated in Los Angeles, with 29 cases referred to Los Angeles County prosecutors, resulting in 14 prosecutions. This is a small but significant dent in the prosecution of worker deaths. More important, it sends a powerful message to employers in Los Angeles County that worker safety violations can result in criminal prosecutions.

On June 11, 2015, the U.S. Attorney for the Eastern District of Pennsylvania indicted the owner of a roofing company in connection with the fatal fall of an employee. James J. McCullagh Roofing was charged with four counts of making false statements, one count of obstruction of justice, and one count of willfully violating an OSHA regulation leading to the death of an employee who was killed after falling approximately forty-five feet from a roof bracket scaffold while performing roofing work. According to the indictment, McCullagh failed to provide fall-protection equipment to his employees. During the OSHA investigation, McCullagh compounded this tragedy by attempting to cover up his failure to provide fall protection when he falsely stated, on four occasions, that he had provided fall-protection equipment, including safety harnesses, to his employees. McCullagh told an OSHA compliance safety and health officer that his employees had been wearing safety harnesses tied off to an anchor point when he saw them earlier in the day prior to the fall. The indictment alleges that McCullagh knew that he had not provided fall protection to his employees and that none of his employees had safety harnesses, or any other form of fall protection. It was further alleged that McCullagh directed other employees to falsely state that they had fall protection, including safety harnesses, on the day of the fall. McCullagh pleaded guilty and was sentenced on March 29, 2016, to ten months in prison and one year of supervised release.[85]

In May 2015, a Santa Clara County jury convicted two corporate officers of U.S.-Sino Investments Inc., a construction company in Southern California, of involuntary manslaughter in the death of a thirty-nine-year-old worker from Mexico. The worker, along with his brother—both day laborers—was hired off the street to work on a hillside house that was behind schedule and seriously over budget. A building inspector, after seeing a trench with no shoring, and after soaking rains had made the hillside unstable, had issued a stop-work order while the contractor consulted with a soil engineer. The stop-work order was ignored, and three days later the trench collapsed around the worker, killing him.[86] In response to the jury verdict, the defense lawyer predictably vowed to appeal, claiming the jurors did not follow the jury instructions. Moreover, in a moment of unintended irony, the lawyer added that the jury verdict should "shock" all CEOs. Yes. That is exactly the point.

Finally, in July 2016, the San Francisco District Attorney's Office charged the owner of a granite supply company with two counts of involuntary manslaughter in the February 2014 deaths of two employees. The workers were crushed to death when the forklift used to unload the slabs of granite tipped over, killing the men. The owner who was operating the forklift failed to assess the hazards of moving the slabs and failed to train the workers.[87]

On April 5, 2010, at 3:02 p.m., an explosion erupted at the Upper Big Branch Mine (UBB) near Montcoal in Raleigh County, West Virginia. Fueled by high levels of coal dust, methane gas exploded in a fireball deep inside the UBB mine and roared through more than two miles of underground tunnels. Twenty-nine miners were killed.

The UBB mine was owned by the Massey Energy Company, which was the fourth-largest coal producer in the United States. At the time of the explosion, Massey's CEO was Don Blankenship, a tall and physically imposing figure, with dark, brooding features. Blankenship rose from poverty in West Virginia to become a powerful and feared coal baron. He used the courts and politicians to shield Massey, and himself, from accountability, while becoming the largest coal producer in central Appalachia. As an example of Blankenship's influence, he contributed $3 million to defeat an incumbent justice on West Virginia's Supreme Court, at a time that Massey was appealing a $50 million civil judgment. Predictably, the court overturned the judgment. The deciding votes were provided by the newly elected Justice Brent Benjamin, whom Blankenship helped elect, as well as the chief justice.[88]

Four months earlier, Blankenship and the chief justice were reportedly seen vacationing together on the French Riviera.[89] In June 2009, the United States Supreme Court reversed the decision based on the undue influence that Blankenship had on Justice Benjamin's campaign for the West Virginia Supreme Court, at a time when Massey had a pending case before it. The U.S. Supreme Court concluded "that Blankenship's campaign efforts had a significant and disproportionate influence in placing Justice Benjamin on the case."[90]

Because of Blankenship's outsized influence, no one believed that he or Massey would be held accountable for the deaths of the twenty-nine miners. Only two years before the explosion at UBB, federal prosecutors refused to bring charges against Blankenship for an explosion at another Massey-owned mine, which killed two miners. But shortly after the UBB explosion, President Obama appointed Booth Goodwin to become the new U.S. Attorney for the Southern District of West Virginia. While other investigations into the UBB disaster continued, including congressional hearings in which mine safety reforms were blocked by House Republicans,

Goodwin's office methodically began to build its case against Massey and Blankenship. The investigation went forward despite a public relations campaign mounted by Massey in the immediate aftermath of the explosion, which warned against a "rush to judgment," and a statement by Blankenship around the time of the funerals that "any suspicion that the mine was improperly operated or illegally operated or anything like that would be unfounded."[91] During the investigation, Blankenship, and other Massey officials, asserted their Fifth Amendment rights in December 2010 and refused to answer government investigators' questions. Two months later, Massey's security chief was charged with lying to investigators and trying to destroy evidence. He was convicted and sentenced to prison. The following year, in February 2012, the UBB mine superintendent was charged by Goodwin's office with conspiracy to violate mine safety laws. The superintendent pleaded guilty and spent twenty-one months in jail. In the meantime, West Virginia legislators passed a mine safety reform bill, greatly watered downed by industry lobbyists, and which prominently included a drug-testing provision for miners, even though there was no connection between drug use and the UBB explosion. Still, the investigation continued. The big break came when Goodwin's office announced that a longtime Massey official, David Hughart, would plead guilty to a conspiracy to violate mine safety laws and would cooperate with the criminal investigation. In entering his plea of guilty, Hughart alleged that Blankenship was part of the conspiracy. Hughart was sentenced to forty-two months in prison. The noose around Blankenship was getting tighter.[92]

Finally, on November 13, 2014, Goodwin announced the indictment of Blankenship.[93] In a forty-three-page criminal indictment styled *United States of America v. Donald L. Blankenship*, Goodwin's office charged Blankenship with four counts of conspiring to violate mine safety laws related to the ventilation and control of coal dust rules, and a conspiracy to cover up the violations, and to interfering with the government's enforcement efforts by providing advance warnings of mine inspections.[94] Reading like a novel of abuse of power, of threats and intimidation, the indictment charged Blankenship "with the routine violation of its [Massey] ventilation plan," and that beginning in April 2009, Blankenship requested and received daily reports detailing Massey's violations of mandatory federal mine standards and estimates of the fines for these violations.[95] In the first six months of 2009, the reports that Blankenship daily received showed that the UBB mines were cited for approximately 596 violations of mine safety laws, mostly related to the accumulation of combustible coal dust that eventually ignited in a fireball.[96] A major cause of the accumulated dust was that Blankenship deliberately understaffed UBB with too few miners dedicated to "non-coal production time," which is necessary to make sure that the safety laws are being followed.[97] During this same period, Blankenship demanded and received production reports (along with any reasons for production delays) every thirty minutes, including at his home on evenings and weekends.[98] The indictment revealed Blankenship's micromanagement of the UBB mine, down to how many miners were hired.[99] Blankenship, the indictment charged, pressured managers to cut the number of miners assigned to critical safety-law compliance jobs, including those jobs required to properly ventilate and clean up accumulated coal dust.[100] He threatened the jobs of mine managers over safety compliance, and tied their compensation to the routine violation of safety laws.[101] The indictment charged Blankenship with instructing coal managers to "run some coal" and that "we'll worry about ventilation and other issues at an appropriate time. Now is not the time."[102] In announcing the indictment, Goodwin took note to say that a conviction carried a maximum prison time of thirty-one years.[103]

In response, Blankenship attacked the indictment as "political," while his lawyer asserted that Blankenship was "a tireless advocate for mine safety."[104] Blankenship retired from Massey in December 2010, in a deal that provided him with $12 million in cash.[105] The following month, Alpha Natural Resources Inc. announced that it would purchase Massey for over $7 billion.[106] In a prepared statement after the indictment, Alpha took pains to note that Blankenship left Massey before the acquisition and that he "was never an employee of Alpha Natural Resources."[107]

A superseding indictment was filed on March 10, 2015, that consolidated some of the conspiracy allegations.[108] Blankenship's lawyers filed numerous motions, including a "Motion to Dismiss for Selective and Vindictive Prosecution." Blankenship claimed that he was being prosecuted in violation of his First Amendment rights only after he released a documentary film titled "Upper Big Branch—Never Again," a piece of YouTube propaganda that blamed regulators for the explosion, and that the film inflamed the "West Virginia Democratic Establishment," including Senator Joe Manchin, as well as union leaders.[109] U.S. District Judge

Irene Berger denied Blankenship's motion.[110] She also denied his motion to disqualify Judge Berger, *"and All the Judges of the United States District Court for the Southern District of West Virginia."* Blankenship's lawyers claimed that Judge Berger and others could not be fair because Goodwin's father is a judge presiding in the same district.[111] This claim came from the same man who the United States Supreme Court held exerted an "exceptional case" of influence, and which created a "probability of bias" that required Justice Benjamin's recusal. Judge Berger set the case for trial in October 2015.[112]

In a not entirely coincidental matter, in April 2015 the Department of Labor reported that black lung disease had killed more than seventy thousand miners between 1970 and 2013, while permanently disabling many thousands more.[113] Black lung was supposed to have been eradicated in 1969 with safety reforms, including regulations on coal dust control and ventilation.[114] At the same time, the Mine Safety and Health Administration reported that miner deaths increased in 2014, totaling forty-four, while a record low of sixteen deaths were reported in coal mines.[115] A couple of thoughts on these facts. Blankenship and Massey were certainly not responsible for all the black lung deaths in this time period. But it is just as certain that at Massey plants, including where the indictment alleges, among other things, a criminal disregard for coal dust and ventilation regulations, many miners there developed black lung. And it is likely that, under Blankenship's reign, many more miners died from the disregard of safety laws at UBB mines than only the twenty-nine miners killed in the UBB explosion. And, finally, the decline in coal miner deaths in 2014 just may have had something to do with Booth Goodwin's four-year investigation, and prosecution of Massey officials, leading up to Blankenship's indictment.

In the end, Don Blankenship finally stood trial. At the Robert C. Byrd United States Courthouse in Charleston, West Virginia, jury selection began on October 1, 2015. According to court transcripts, eighty-eight citizens living within the twenty counties that make up the Southern District of West Virginia were summoned for jury duty. After the first day of jury selection, forty-two potential jurors were dismissed, whittled down through *voire dire* examination by Judge Berger, and from answers on questionnaires originally mailed to over three hundred potential jurors in the district. Potential jurors were excused for cause, including one who worried that Blankenship might be a "scapegoat," while another said that Blankenship was "a horrible person with questionable morals and no work ethics."[116] Jury selection lasted four days before twelve jurors and alternates were finally seated. The evidentiary portion of the trial lasted twenty-four days.[117] The government called twenty-seven witnesses, including miners and UBB officials who testified that safety violations were a known and accepted way of doing business. More than five hundred exhibits were introduced into evidence. The defense rested without calling a single witness, including Don Blankenship.[118]

The jury deliberated for parts of ten days, despite defense motions for a mistrial. On December 3, more than five years after the UBB explosion, the jury convicted Blankenship of criminal conspiracy to willfully violate federal mine safety standards, convinced beyond a reasonable doubt that he ruthlessly demanded that miners produce as much coal as possible and without regard to their safety. He was acquitted on the lesser charges of securities fraud and making false statements to investigators.[119]

While Booth Goodwin and his team achieved a conviction on the most important count in the indictment in terms of the safety of miners at UBB and elsewhere, the crimes of securities fraud and lying to investigators involved more jail time—possibly thirty years in prison—than knowingly violating thousands of safety rules at UBB, which ultimately resulted in the explosion of a fireball killing twenty-nine miners. In a final insult to the families of the dead miners, the count of conspiracy to willfully violate federal mine safety standards is only a misdemeanor, with a maximum one-year prison sentence and a $250,000 fine. In the criminal justice system, jurors are not told by the court during deliberations what the penalty is for a conviction of any count.[120] Jurors decide guilt or innocence based on the evidence, while Congress decides the penalty. According to the wisdom of Congress, known violations of mine safety laws that resulted in the deaths of one or twenty-nine miners are punishable by only up to one year in prison.

Blankenship's lawyers immediately filed an appeal, claiming that the trial was "full of errors," and the lead lawyer boldly declared that he was "as confident as I've ever been" that the verdict would be thrown out by the Court of Appeals.[121] In an interview after the trial, one of the jurors expressed frustration at learning that Blankenship was facing only a possible one-year sentence.[122] Bill Rose, who drove 118 miles each day to the courthouse, said, "I wish we could have got more, you know—because I know—I assume since the other ones were felonies—he would ultimately get more fines or jail time, but that's not how it went down."[123]

In the immediate aftermath of the verdict, the fault lines cleaved along the issue of the need for more oversight and regulation, and stronger criminal penalties.[124] Leading the charge against the prosecution, the *Wall Street Journal* naturally trotted out its favorite bogeyman and wrote that turning an industrial accident into a crime was yet another example of an "Obama-era" prosecution. The *WSJ* declared victory for the defense as Blankenship avoided the prospect of more serious jail time, and audaciously asserted that "Massey had paid dearly" in the millions of dollars that it paid in fines and settlements, while giving short shrift to the suffering of the twenty-nine families and the Big Branch community.[125]

A day after the verdict, U.S. Representative Bobby Scott, a Virginia Democrat, called for tougher penalties for mine violators who knowingly ignore mine safety standards that expose miners to significant risk of injury, illness, or death.[126] The Robert C. Byrd Mine Safety Protection Act, reintroduced on April 30, 2015, in the House by Scott and in the Senate by Bob Casey, a Pennsylvania Democrat, would make it a felony punishable by up to five years in prison for doing what Blankenship was convicted of by the Charleston jury.[127] There were no Republican cosponsors of the legislation, including any of West Virginia's three Republican House members. Govtrack.com gave the bill a 2 percent chance of passage.[128]

In the meantime, Blankenship sought and received permission from the court to leave the Charleston jurisdiction for the Christmas holidays in 2015 to travel back to his new home in Las Vegas.[129] For the families of the dead miners, and other miners hoping for an effective deterrent to safety violators, all they received was a lump of coal.

On May 12, 2016, Don Blankenship surrendered to authorities to begin serving his one-year sentence at the Federal Correctional Institution Taft near Bakersfield, California. Blankenship became a federal prison inmate around eight hours after a three-judge panel of the Fourth Circuit Court of Appeals denied his motion for a stay of his prison sentence while his appeal was pending. The circuit court's brief order simply stated: "Upon review of submissions relative to the motion for release pending appeal, the court denies the motion." Before turning himself over, Blankenship issued a letter to his supporters that thanked them for giving him the strength "to endure an indescribable injustice." Patty Quarles, the mother of Gary Wayne Quarles, who died in the UBB explosion, simply said in response to the news of Blankenship's incarceration, "I wish it was a lot longer, but we will take this."[130]

Meaningful prison time measured in years, and not months, is a deterrent. Is it the silver bullet? Certainly not. But, it is, and must be, an integral part of comprehensive occupational safety and health reform. There must be a very high price to pay for key decision makers when worker safety rules are ignored and violated. So, to all the Don Blankenships who daily perform cost-benefit analyses on compliance with worker safety laws versus corporate bottom-line profits and shareholder return, it must unequivocally be declared that crime against workers does not pay.

Workers' Compensation Reform: In some variation it has been said that a nation's greatness is measured in how it cares for its most vulnerable citizens. This sentiment is most often expressed in relation to how we treat the elderly, the homeless, disabled persons, and our children. Missing from any list that I have seen are the injured worker and his or her family—and, too often, the surviving family members of a worker killed on the job. Just how vulnerable is this class of our citizens, and how do we care for them? The short answers are: very vulnerable, and not very well. The long answers are more complicated.

When a worker gets injured or killed on the job, there is a patchwork of flinty systems for financial and medical compensation that the workers and their families must navigate, usually alone, and at a time when they are physically, and otherwise, suffering the most. These workers often do not possess the requisite knowledge, skill, and resources to optimize their receipt of critical medical and income replacement benefits available to them, which in turn leads to further bad outcomes.

The two main benefit systems in play for injured workers are state workers' compensation and federal Social Security Disability Insurance (SSDI). State workers' compensation programs provide nearly $60 billion per year in cash and medical benefits, while SSDI has annual cash outlays to disabled workers approaching $100 billion. These amounts are not cumulative but have statutory offsets that limit the amount paid to an injured worker receiving up to a maximum of 80 percent of their pre-injury average earnings. There is some overlap among the two systems for workers who become disabled, since a substantial proportion of disability in the United States is caused by work-related injuries and illnesses.[131] Yet, as a safety net for injured workers and their families, these two systems fail to adequately compensate injured workers at a time when they need it the most.

For one, SSDI is available only if a worker has worked long enough, and recently, and his or her employer has paid into Social Security on the worker's behalf. Putting aside the substantive legal issues of whether a worker is disabled and unable to work, there is the matter that a worker has to be in the Social Security system in the first place. As a result, for the millions of workers who are paid in cash, or unlawfully misclassified as independent contractors, and thus kept off the books by employers skirting state and federal tax laws and other payroll obligations, and who often work in some of the most dangerous jobs, SSDI is simply of no avail to them. Most important, because workers' compensation benefits provide only a fraction of long-term lost earnings for disabled workers, SSDI has become the default insurance that reduces "the financial burden of the long-term consequences of workplace injuries on the most severely disabled." While it can be a good outcome for those workers who can take advantage of the benefits of SSDI, this cost shifting for long-term work-related disabilities undermines the incentives for employers to provide a safe workplace. Accordingly, the market incentives for safe workplaces through higher workers' compensation premiums are inadequate as a result of this cost shifting, and "the scale of the inadequacy is potentially quite large."[132]

The current workers' compensation system is broken, not only for injured workers, but for workers not yet injured but nevertheless at substantial risk for injury, owing in part to the disincentives undermining workplace safety. But before we can fix the system, let us first understand its original promise.

The history of workers' compensation traces its roots all the way back to approximately 2050 BCE and the Mesopotamian fertile crescent of ancient Sumeria.[133] The law of Ur, found written on stone tablets, provided financial compensation for specific injuries to workers and their body parts, including for fractures. Similar compensation schedules for the loss of a body part from work followed in the Code of Hammurabi, as well as with the ancient Greeks, Romans, Arabs, and Chinese. Every advanced society codified detailed "schedules" to compensate workers for their injuries. In Europe, during the Middle Ages, the compensation schedules of ancient civilizations were replaced, and serfs became beholden to the arbitrary benevolence of their feudal lords to take care of them when they became hurt. Eventually, as English common law developed, a jurisprudence governing worker injuries developed as well, and with it, three principal defenses to a worker's claim for compensation.[134]

The doctrine of contributory negligence shielded an employer from liability if the worker was in any way at fault. In the United States, this was established in a case where a freight conductor fell from his train. Although a loose handrail was blamed, the conductor was not entitled to any compensation because he was at fault for not checking for faulty equipment. The fellow-servant rule further protected employers from liability if a worker's injuries were in any way the fault of a fellow employee. In both these doctrines, there was no allocation of liability to an employer. It was an absolute defense. Finally, there was the assumption-of-risk doctrine, which held that workers knew the risks of their jobs. This principle was often codified in contracts that were little more than waivers by workers of their right to sue for a work-related injury. These contracts were known as the "worker's right to die," or "death contracts."[135] Even with these judicially created limitations of liability, workers still used the courts as their only means of financial recompense for their injuries by suing their employers under tort laws in jury trials. Eventually, the unpredictability of juries, and the high costs of jury verdicts, had businesses clamoring for a change in the legal status quo.

The modern workers' compensation system was born in Germany at the end of the nineteenth century under the leadership of Chancellor Otto von Bismarck. He was not a workers' advocate, but a consummate politician. In response to the growing strength of workers' political parties in Western Europe, most notably the Social Democratic Party, the Workers' Accident Insurance (1884) was enacted under Bismarck, as well as public pensions and other forms of public aid. While providing medical and rehabilitative benefits to injured workers, the Workers' Accident Insurance was a worker's exclusive remedy, thereby barring employees from suing their employer in the civil courts.[136]

In the United States, the "grand bargain" between workers and their employers, which was initially opposed by the fledgling labor movement, would have to wait until 1911, when Wisconsin became the first state to establish a workers' compensation system. By 1920, most states established similar programs, with the last holdout being Mississippi in 1948.[137] While each state's workers' compensation system varied in one degree or another, depending on legislation and court rulings, all shared the essential framework that made the "grand bargain" possible. Employers were required to cover medical costs, and partial wage replacement, that "arise out of or in the course of employment." In exchange, employers were granted immunity from lawsuits and civil liability for injuries to their employees. With few exceptions, like willful and gross negligence, this immunity has remained unchanged today.

Except for Texas since 1913, and Oklahoma in 2013, where employers can opt out of workers' compensation coverage and lose their immunity from civil lawsuits, most employers are required to provide coverage for their employees' work-related injuries. Mandatory coverage has been funded by the purchase of private insurance. Each employer's premium is rated based on a complex and wide range of experience factors related to an employer's industry and risk.[138] A roofing contractor will generally have a higher premium than a bank or accounting firm. And there are incentives for employers who enact safety programs designed to reduce worker injuries. But, as well intended as it was at the outset, over the years this "grand bargain" is no longer grand or a bargain for workers or American taxpayers.

Examining the costs of workplace injuries, the National Safety Council estimated that in 2012 the cost of fatal and nonfatal work injuries was $198 billion, a number on par with the cost of other common and debilitating illnesses such as diabetes and dementia, which were estimated at $245 billion, and $159 billion to $215 billion, respectively.[139] These and other medical illnesses and conditions are covered by private insurance now mandated and subsidized (at this writing) by Obamacare, or state and federal Medicaid and Medicare programs. However, in recent years, when it comes to workers' compensation coverage, which was originally intended to provide insurance for lost wages and first-dollar medical coverage, state legislatures have enacted reductions in workers' compensation benefits in their efforts to balance state budgets and impose austerity measures on the backs of workers, while promising a pro-business climate in order to attract new companies to their states.

In Alabama, for example, workers' compensation employer costs per $100 of workers' wages declined from $3.18 in 1988 to $1.81 in 2014.

In Massachusetts, which created the first labor bureau in 1869, average premium costs for employers declined from $3.67 in 1988 to $1.17 in 2014.[140] Similar declines have occurred almost everywhere in the United States, to the point where the average premium cost to employers per $100 of workers' wages in 2014 was its lowest ever, at $1.85.[141] As a result, who pays the annual bill for workers' injuries, which in 2012 nearly topped $200 billion, has shifted dramatically. In 2015, OSHA reported that injured workers pay 50 percent of the costs in the form of out-of-pocket expenses. Government programs and private health insurance pay a combined 31 percent of the costs. At the bottom are employers and their workers' compensation insurance carriers. They pay only 21 percent of the bill.[142] Moreover, OSHA reported that fewer than 40 percent of eligible workers apply for any workers' compensation benefits, while as many as 97 percent of workers with occupational illnesses receive no benefits at all.[143] Added to this is the fact that hundreds of thousands of workers are misclassified as independent contractors and receive no workers' compensation coverage whatsoever from their contracting employer.[144] According to the OSHA report, in Texas, North Carolina, and Florida alone, more than half a million construction workers were misclassified as independent contractors and as a result were not eligible for workers' compensation benefits. Their misclassification cost taxpayers over $2 billion a year in lost tax revenues, while shifting medical costs back onto the same taxpayers.[145] Moreover, employers who misclassify do not worry about providing a safe workplace, since OSHA and state laws do not reach the self-employed. Accordingly, these employees are more likely to get injured and less likely to be covered by workers' compensation. Finally, temporary workers often do not report their work injuries for fear of being classified as troublemakers and not being assigned to temporary work assignments.

In the end, the cost shifting, declining benefits, and lack of coverage are exacting an economic toll on injured workers and their families that goes far beyond the immediate injury. Studies have reported that injured workers lose an average of 15 percent of their earning potential in the ten years following an injury, with their incomes on average being $31,000 lower over ten years than if they had not been injured.[146] For workers, especially low-wage workers, who struggle from paycheck to paycheck, the loss of income from a primary wage earner is especially devastating and creates a cycle of poverty from which a worker and his or her family are likely never to recover. This in turn creates foreseeable collateral damage in the form of low self-esteem, and stress on families already under pressure to make ends meet. Partners or spouses who have to care for an injured wage earner are often forced to reduce their hours, or change jobs, which only further exacerbates their financial hardships.

However, instead of fixing the broken workers' compensation system, a consortium of more than two dozen major corporations, including Walmart, Safeway, Lowe's, Macy's, Kohl's, and Nordstrom, have been funding a national campaign to replace the existing workers' compensation system with an even more onerous system for injured workers. The campaign's lobbying arm, the Association for Responsible Alternatives to Workers' Compensation (ARAWC), has been intent on rewriting the existing laws by, among other things, allowing companies to opt out of workers' compensation coverage and permitting them to provide coverage under their "own rules governing when, for how long, and for which reasons an injured employee can access medical benefits and wages." Not surprisingly, ARAWC's CEO is also Walmart's head of risk management. Both Texas and Oklahoma already have opt-out provisions in their laws, which cap benefits and in Oklahoma require employees to report injuries before the end of their shift or lose their eligibility for coverage. ARAWC has spent thousands of dollars to lobby legislators in other states, including Tennessee, where a proposed bill would not require employers to pay for artificial limbs, hearing aids, home care, funeral expenses, or certain disability benefits, all of which are required by current laws.[147]

In addition to the threat from ARAWC, other national trends in workers' compensation hammer away at the "grand bargain." These include the right of employers to choose an employee's doctor, who together with the insurance adjusters makes the medical decisions for the injured worker. In 2015, there were thirty-seven states where workers cannot choose their own doctor, or are limited to a list of preferred medical providers—preferred, that is, by their employers.[148] As a result, these decisions are more often about the best interests of the employer and its insurance company than about the well-being of injured workers.

One of the most insidious developments is the requirement that employees submit to drug and alcohol testing after a job injury, regardless of whether there is any evidence, or belief, that drugs or alcohol played any part in the work injury. This is even so when the accident is caused by another employee. In this regard, it is well established that the most popular test for marijuana use, a urine test, is also the most unreliable to detect contemporaneous use. For even an occasional user, the metabolite THC can remain in the system for days after use, and weeks for a regular user. Thus, the obvious intent of post-accident testing is to reduce the likelihood that an employee will report a compensable injury, for fear of testing positive and risking termination, even without proof that they were impaired when the injury occurred.

Finally, a number of studies found that as many as 50 percent of workers underreport work-related injuries that are serious enough to require medical attention. Among the reasons for underreporting is the fear of retaliation. Retaliation, and other fears, including drug and alcohol testing, and threats of discipline against workers for "unsafe behaviors," all serve to chill an injured employee's right to compensation. In 2012, OSHA issued a memorandum condemning employer practices that dissuaded employees from reporting injuries. One of the practices condemned was financial incentive programs that reward employees for not reporting injuries. OSHA concluded that such programs interfere with the right of employees to report workplace injuries and in turn result in underreporting by both the employee and the employer.[149]

Clearly, the "grand bargain" is in tatters. What was originally the first social program for workers has become a commodified for-profit system, rife with conflicts of interest among doctors, insurance companies, and employers. The medical and financial interests of injured workers and their families have been relegated to something that must be challenged and avoided. What can be done to restore the original promise for workers that dates back to ancient times? The following is a wish list of real reforms that have been advanced by workers' rights advocates, which if all, or in some combination, are adopted would go a long way to restoring the original promise of workers' compensation.

First, workers need to be better informed about their rights to workers' compensation benefits. Access to benefits should not be something feared or unknown, but a right of every worker. Every employer must be obligated to clearly inform its workers of their rights to workers' compensation benefits. This is especially true given the lack of uniformity in state workers' compensation laws at a time of greater worker mobility across state lines. Access should not be limited by a lack of information. President Theodore Roosevelt understood this when, in January 1908, he said in a special message to Congress after the Supreme Court found unconstitutional the Employers' Liability Act of 1906: "It is a matter of humiliation to the Nation that there should not be on our statute books provision to meet and partially to atone for cruel misfortune when it comes upon a man through no fault of his own while faithfully serving the public. In no other prominent industrial country in the world could such gross injustice occur."[150]

Second, the receipt of medical benefits should not be determined by doctors beholden to insurance adjusters and to their employer clients. As it now exists, the workers' compensation system in most states has created an inherent conflict between patients and medical providers that arguably violates the latter's Hippocratic Oath and, as important, the American Medical Association's Code of Ethics with respect to doctors who perform work-related and independent medical examinations. The latter requires that a doctor "evaluate objectively the patient's health or disability. In order to maintain objectivity, IEPs [industry employed physicians] and IMEs [independent medical examiners] should not be influenced by the preferences of the patient-employee, employer, or insurance company when making a diagnosis during a work-related or independent medical examination."[151] In addition, the AMA code requires doctors to fully disclose "potential or perceived conflicts of interest" that may arise, for example, from a doctor's relationship with an insurance company or employer. However, in many states, injured workers have to go to company doctors for treatment, or pay for their medical care themselves. Additionally, the medical opinion of a worker's doctor is frequently challenged by the insurance companies' doctors, setting up a drawn-out and costly battle of "independent medical exams" and conflicting medical opinions. This broken system must be replaced by a system where every injured worker receives truly independent medical care from a provider whose only interest is to heal workers and return them to work, if possible.

Third, workers' compensation medical benefits should be integrated into a single-payer national health care system instead of pitting states against each other in a race to the bottom. Instead of insurance premiums, which imbed high administrative costs in for-profit insurance companies, a single-payer system can be publicly funded by taxes, similar to Medicare, along with penalties and incentives for employers, depending on how unsafe or safe their workplace is. In a national system, benefit schedules would be standardized. Currently, where you work and get injured matters. As a result, there are widely divergent schedules of benefits for injured body parts that make little economic sense or good public policy. The loss of an arm in Alabama, for example, is worth up to $48,840. An arm in Ohio is worth $193,950, and $439,858 in Illinois. The loss of a big toe ranges from $6,090 in California to $90,401 in Oregon.[152] We can learn from the ancient Sumerians.

Fourth, simplify and streamline the dispute resolution process when there is a legitimate legal issue about a worker's claim. Too often insurance companies deny claims in bad faith, hoping to use that as leverage to force workers to settle for less than they would otherwise be entitled to. When this occurs, insurance companies should be required to pay substantial penalties for the bad-faith denial of a claim, and for recidivist insurers their license should be suspended or revoked. But what is really needed is a complete overhaul of the workers' compensation legal system. What was initially supposed to be a "no fault" system, expressly protecting employers from lawsuits in exchange for the payment of medical and income benefits, has become a complex and oppressive legal system that requires employees to bear the burden of establishing their entitlement to benefits. The original premise was brilliant in its simplicity; if an employee was hurt at work, that employee was entitled to compensation. Now, insurance companies routinely throw up legal challenges to an injured worker's claim, beginning with whether the injury was work related. This is an especially difficult obstacle for the millions of workers who do not suffer an acute injury, such as a broken leg from a fall, but instead are injured over time, as occurs with musculoskeletal injuries, and those who become sick from exposure to chemicals or other toxic substances in their workplace. Proof of a work-related injury in these cases often requires expert testimony, which is very expensive to obtain and beyond the financial means of most workers. In addition, a myriad of other procedural hurdles are placed in the way of injured workers, including whether the employee reported the injury within the narrow time limits set out in most state statutes. Again, for workers who have not suffered an acute injury, they may be precluded from obtaining compensation for injuries that manifest themselves over months and years. In the end, the current legal system is designed to make it more difficult, not less, for workers to obtain their rightful benefits.

Fifth, statutory attorney's fees for a worker's attorney have been eliminated or reduced in many states. This was done to make it more difficult for an injured worker to obtain benefits when they are denied. Attorney's fees should be returned to rates that fairly compensate attorneys who represent injured workers in their claims for medical benefits and income replacement. The truth is, if the system operated as intended, there would be little need for attorneys. But until the system is truly reformed, the legal playing field must be level for injured workers.

In the end, until real reforms such as these are implemented, the current system, which today still creates the same gross injustice to American workers that Teddy Roosevelt railed against more than a hundred years ago, will continue in crisis. And in the end, America will have failed some of its most vulnerable and important citizens—its workers.

Regulatory Reform: It is generally a good day for government and the public body when Congress enacts new legislation. Almost all Americans will come into contact at some point in their lives with the Social Security and Medicare systems, and except for fringe outliers, most would agree that these historic legislative achievements have been good for all Americans. Beyond Social Security and Medicare, there are hundreds of other pieces of legislation creating federal agencies that daily impact all our lives, and of which there is no serious dispute as to their efficacy. Think of the Federal Highway Administration, the Federal Aviation Administration, the Food and Drug Administration, the Transportation Security Administration, and many, many more. Then there are agencies and laws that arouse an entire range of emotions, including controversy, unwavering support, and outright enmity. In this era of hypercharged partisanship, the most controversial legislation would notably include the Affordable Care Act (Obamacare) and the Dodd-Frank Wall Street Reform and Consumer Protection Act—but also such landmark achievements as the Civil Rights Act of 1964 and the Voting Rights Act of 1965. Then, somewhere in the middle are an alphabet stew of laws and agencies that most Americans pay little attention to but are the focus of pitched battles between regulators and public interest groups on the one side, and armies of lawyers and lobbyists representing elite special interest groups on the other. Clean air and water legislation under the jurisdiction of the Environmental Protection Agency come to mind, as do, of course, OSHA and the OSH Act.

What all these laws and agencies have in common is that they are generally not self-enforcing based on the plain language of the statute. Rather, each has been authorized by Congress, to lesser and greater degrees, to issue regulations and rules to solve the problems and achieve the goals underlying the legislation. Take the OSH Act, for example. The legislative goal, as announced by Congress in the General Duty Clause, requires that "each employer shall furnish to each of his employees employment and a place of employment which are free from recognized hazards that are causing or are likely to cause death or serious physical harm to his employees."[153]

In line with this seemingly unequivocal declaration, it is unthinkable that anyone could seriously argue that worker health and safety is not an important area for Congress to regulate. Who would stand against protecting American workers from harm in the workplace? The simple answer is, no one. However, agreeing that workers should not be harmed in the workplace, and implementing this legislative goal, are two completely different matters. That's where the regulatory rubber meets the road. Until rules or regulations are proposed, publicized, and put into final form, a law that Congress passes is often just a bunch of words.

The New Deal was the single most transformative period of government since the founding of the Republic. The federal government's role in American life exploded in response to the catastrophe that was the Great Depression. Regulatory government was born out of the ashes of the Great Depression and Franklin Roosevelt's New Deal policies. Eighteen new agencies were established, more than were created at any time before 1900. The rolls of government employees swelled as the regulation of commerce and private conduct became a matter for the federal government in Washington. In response to this enormous growth in government, and after years of studies, reports, and political fighting, the Administrative Procedures Act (APA) was enacted in June 1946 in an effort to regulate the regulators by establishing uniform standards for rule making and the adjudication of agency actions.[154]

Sometimes referred to as the "fourth branch" of government, or the "modern administrative state," the decisions of the regulatory agencies "dwarf the other three branches, certainly by volume and arguably by importance, as well. Simply put, modern government *is* administrative government. [. . .] From the food and water citizens ingest to the air they breathe, every aspect of modern life is thoroughly shaped by the decisions of regulatory agencies."[155]

On the rule-making side, the standards seem simple enough. The Office of the Federal Register has a publication that succinctly lays out, step by step, how a rule or regulation comes into being. It is almost as simple as following a cake recipe.[156]

The agency charged with implementing the legislation will develop a new rule or regulation that falls within its statutory jurisdiction. The new regulation may be proposed in response to external factors, such as concerns at the Department of Labor over an increase in workplace injuries or illnesses, or at the FDA from E. coli outbreaks. Or, in the case of new legislation, regulations are proposed to guide the stakeholders and the public as to the intent and scope of the legislation. Each agency proposes these rules based on its own special expertise from the researchers, economists, scientists, lawyers, and analysts it employs. After identifying a problem, these experts will spend many months gathering information in the pre-rule-making stage, known as the Notice of Proposed Rulemaking, before eventually publishing a proposed rule for public comment. The proposed rule is published, not in an easily accessible format for the everyday American, but in the *Federal Register*, an incredibly dense and arcane daily publication of the activities of the federal government that only the wonkiest wonks in a particular area will bother to read. Reading the *Federal Register*, including notices of proposed agency rules, is not something a person would casually do over a mocha latte at Starbucks. But it is the reading material that informs the public about what the federal government is doing, and which asks for its input. After receiving public comments, the agency may proceed with revising and tweaking a final rule. At the last step, the final rule is published, and it has the force of law.

To be sure, there are other steps that must be followed, such as approval in the first instance for proposing a new rule by the Office of Information and Regulatory Affairs (OIRA) at the White House. OIRA can grant permission to proceed with a proposed rule, or refuse permission based on the potential "significant" economic impact, or important policy considerations that it addresses. As such, many proposed rules may never see the light of day because of the concerns of the then current White

House resident's political fortunes. And even if a rule becomes final, Congress may still have the last word, and it can disapprove and block a final rule from taking effect. Finally, if a rule has passed all these hurdles, it can be challenged and delayed in court by a party that claims that it has been damaged or adversely affected by the rule, or that the agency violated the APA in the process. In the end, this seemingly simple and transparent process can easily and willfully be turned into regulatory quicksand by disabling the goals that the enabling legislation sought to promote while leaving millions of Americans at risk and vulnerable.

In reaction to the New Deal and programs like Social Security, the modern antigovernment movement was born. However, because of the enormous popularity of these programs, and fearful memories of the Great Depression still fresh in people's minds, the antigovernment movement in the postwar years was largely relegated to the intellectual commentariat of leading conservatives like William Buckley, Ayn Rand, and Norman Podhoretz. There was also Fred Koch, David and Charles Koch's father and one of the founders of the John Birch Society. But by the late 1970s, a sea change occurred, mostly brought on by exploiting the economic insecurities from periods of high inflation, stagnation, and gas shortages, as well as the insecurities of southern whites disaffected by the belief that civil rights policies championed by Democrats unfairly favored minorities at their expense. In 1980, Ronald Reagan was elected, and he wasted no time in championing the cause of the antigovernment movement. In his first inaugural address he famously declared: "In this present crisis, government is not the solution to our problem; government is the problem."[157]

Since then, there has been a titanic struggle over the role of regulatory government among political, media, and corporate elites on the one side, and public-interest groups on the other. Infused into this debate are the catchphrases of the antigovernment groups, which have become an indispensable part of the political lexicon in America to describe all that is wrong with the modern liberal state. These pejorative descriptions such as "big government" and "bureaucratic red tape" form the linguistic divide between those who believe that government exists to promote the common good, or public interest, and those who view government as the obstacle to all economic prosperity, especially for those at the tip of the pyramid.[158]

Today, the antigovernment groups, elevating Ronald Reagan to saint-like status, have attacked regulations of all kind, and for all purposes. Grover Norquist, one of the leaders of the conservative movement, clearly described its agenda when he stated his goal was to cut government in half in twenty-five years, "down to the size where we can drown it in the bathtub."[159] This charge has been funded by the Koch brothers and other similar-minded billionaires, who have unapologetically waged war on the modern administrative state. The weapons at their disposal have indispensably included 24/7 messaging provided by Rupert Murdoch's News Corporation media empire, including Fox News and the *Wall Street Journal*. But their most important soldiers have been the politicians, elected by billions of dollars in campaign contributions.

This relationship, or exchange between moneyed special interests and elected officials, has been described by Steven Croley as the "public choice" theory of regulation, and it goes like this: "Elected politicians prefer to remain in office. They are able to remain in office only so long as they continue to win the electoral favor of their constituencies. Winning that support requires substantial political resources, in particular, votes and money to attract votes during political campaigns. . . . Interest groups possess the very resources politicians require for their political survival."[160] To this exchange, Croley adds a third participant: agency officials who seek to curry favorable budgetary and statutory treatment from legislators, which in turn "motivates agencies to supply desired regulatory outcomes."[161]

By way of example, in September and October 2008, the United States was facing a financial crisis that the chairman of the Federal Reserve at the time, Ben Bernanke, later acknowledged was worse than the Great Depression.[162] Of the thirteen most important financial institutions, twelve were at risk of failure. In the end, the government bailout and rescue of a U.S. financial industry that had self-imploded and brought the nation and the world to the brink of financial ruin are well documented. In its aftermath, an angry electorate demanded accountability and action in exchange for the use of taxpayer monies.

With one exception, no one went to jail for the financial misconduct of Wall Street, despite calls for criminal prosecutions. Congress stepped up with financial reform to make sure that the kind of wild and risky financial speculation that caused the calamity would be forever reined in. Congress passed the Dodd-Frank Wall Street Reform and Consumer Protection Act in 2010, and it was signed by President Obama on June 21, 2010.[163] A summary of Dodd-Frank published by the Senate Banking Committee wrote that its purpose is to "Create a Sound Economic Foundation to Grow Jobs, Protect Consumers, Rein in Wall Street and Big Bonuses, End Bailouts and Too Big to Fail, Prevent Another Financial Crisis."[164] But despite this laudable and unmistakable commitment by Congress, the regulatory promise of Dodd-Frank remains blocked by the money and influence of Wall Street.

Even before the bill was finally passed, the financial industry spent over $1 billion to beat back Dodd-Frank and other related lobbying efforts. As a result, the final version of Dodd-Frank was watered down from its original pledge by, among other things, retaining the sacred cow of "too big to fail" and guaranteeing with a near certainty that taxpayer money will be used to bail out Wall Street at the next crisis.[165] But the financial industry didn't stop at merely influencing the legislation.

Since 2010, the hundreds of rules that are necessary to rein in Wall Street, protect consumers, and prevent another financial crisis have been delayed by the inexorable and relentless efforts of the financial industry. By the close of 2014, Wall Street banks and other financial interests spent $1.2 billion on campaign contributions and lobbying, besting the 2010 number.[166] The return on their investment has been spectacular. As of July 2016, six years after the passage of Dodd-Frank, 271 rule-making deadlines have passed out of 390 total rule-making requirements.[167]

So while in a crisis Congress can pretend to respond to the anger and outrage of Main Street with the kind of vacuous promises of reform to get them through the next election, regulatory delay is the dirty secret that keeps ordinary Americans unsafe and at risk for financial, environmental, and health catastrophes.

In the arena of workplace health and safety regulations, the U.S. Chamber of Commerce and its political allies have pushed their antiregulatory "reform" message that puts American workers in harm's way. The Chamber's antiregulatory misinformation campaign is epitomized through its "This Way to Jobs" campaign. Recycling old scare-mongering, the Chamber has declared, "Congress: Let's build a clear path to job creation and eliminate the cumulative job-killing impact of overregulation before it's too late."[168] Out of the public spotlight, regulatory reform as envisioned by the Chamber will enshrine and codify regulatory delay.

To be sure, there are regulatory outcomes that belie this false narrative that regulatory reform, of the kind that opposes or delays regulations, is a jobs protector. However, the conditions of regulatory success require, in no small part, the cohesive support of elected political elites who can act independent of the moneyed interests. And, in the post–*Citizens United*[169] era of unrestricted money influencing the political landscape and the balance of power, the headwinds of antiregulatory fervor are blowing stronger than ever before. During the Obama administration, regulatory agencies for the most part attempted to promulgate rules that provided benefits to the larger society, at the expense of the powerful few special interests. The Department of Labor was initially headed by Hilda Solis, a liberal congresswoman from Los Angeles, with deep connections to the union movement, and later by Tom Perez, formerly the nation's top civil rights lawyer, and a former aide to Ted Kennedy. President Obama appointed Dr. David Michaels as the head of OSHA. Michaels is an epidemiologist from George Washington University who during the Clinton administration was the chief architect of the initiative to compensate nuclear energy workers for exposure to radiation, beryllium, and other hazards. His assistant secretary and his chief of staff were health and safety directors from national unions. Yet even with agency leadership nominated from these ranks, regulatory action during this time period was constantly under attack from a powerful and well-funded army of corporate lobbyists and lawyers. Even if a rule managed to make it out of the rule-making process, it was attacked in the courts and in Congress itself.

For American workers, regulatory delay has been a constant counterpoint to the promise of OSH Act's General Duty Clause. The following are some of the most egregious examples of regulatory delay.

Crystalline silica is an important industrial material found naturally in the earth's crust. It occurs in many forms, quartz being the most common. As such, it is a common and essential component of sand, stone, concrete, brick, block, and mortar. These materials are used every day in a wide variety of industrial settings, including construction, mining, manufacturing, maritime industry, and agriculture.

Workers in these industries, as a result, are exposed to crystalline silica while performing everyday operations and tasks involving cutting, sawing, drilling, and crushing concrete, brick, block, rock, and stone products. Operations using sand-based products, such as glass manufacturing, foundries, and sandblasting, expose workers to the inhalation of crystalline silica particles in the air. These types of exposures can lead to the development of disabling and sometimes fatal lung diseases, including silicosis and lung cancer. Manufacturing processes with high rates of silicosis include sandblasting, sand-casting foundry operations, mining, tunneling, cement cutting and demolition, masonry work, and granite cutting.[170]

In the United States alone, more than two million workers are exposed to the deadly silica dust. According to the National Institute for Occupational Safety and Health (NIOSH), at least 1.7 million American workers are exposed to respirable crystalline silica, leading to silicosis, lung cancer, pulmonary tuberculosis, and other airway diseases.[171] From 2001 to 2010, an average of 143 workers died each year from silicosis, with an estimate of 3,600 to 7,000 new cases each year.[172] OSHA recently put the exposure number at more than 2.3 million.[173]

In 1971, OSHA adopted a silica standard. It is out of date, and the time for a new standard has long passed. In fact, OSHA began working on a new standard in 1974 after NIOSH recommended that the exposure standards were too high to protect workers from silicosis. Still, rule making on a new standard waited until 1997, in the second term of the Clinton administration. The rule making continued until 2003, when preoccupation with a war in Iraq and terrorism brought rule making to a standstill. In 2009, under the Obama Department of Labor, rule making resumed, until the 2010 midterm elections and the takeover by Republicans in the 112th Congress. On September 12, 2013, OSHA issued a final proposed silica rule sixteen years after rule making began. OSHA received over two thousand public comments on the proposed rule, totaling around thirty-four thousand pages in the public record. OSHA held fourteen days of public hearings, during which all the imaginable stakeholders from more than seventy organization presented testimony. The transcripts from these hearing totaled over forty-four hundred pages.[174]

In the spring of 2015, OSHA released its annual "unified agenda," which was an ambitious list of twenty-three regulations designed to protect American workers. It was ambitious for its scope and breadth, but also because it was a long-overdue to-do list of unfinished health and safety regulations. Among the regulations on the list was the silica rule that had remained stuck in the pre-rule stage ever since it was proposed as a final rule in September 2013. As part of the year-end budget negotiations in 2015, there were threats of a government shutdown, and Senator John Hoeven, a Republican from North Dakota, attached an amendment to the budget specifically prohibiting the use of funds to "promulgate or implement any rule, standard, or policy . . . related to occupational exposure to respirable crystalline silica" until additional studies were completed.[175]

Senator Hoeven's amendment was hardly surprising, since according to the Center for Responsive Politics, from 2009 to 2016, when he received nearly $7,598,605 in campaign contributions, his top industry contributor was the oil and gas industry.[176] North Dakota has been in the midst of an oil boom since the mid-2000s. Silica sand is used in hydraulic fracturing, or fracking. The North Dakota Energy Forum, an industry-supported organization, reported that fracking and horizontal drilling have led to North Dakota's oil and gas boom.[177] OSHA and NIOSH have reported that oil and gas workers are exposed to dangerous levels of silica, especially during hydraulic fracturing.[178] Connect the dots.

As part of the budget negotiations, Republican-controlled congressional committees in both the Senate and the House approved budget cuts of 5 percent for OSHA in fiscal year 2016. At the eleventh hour, and as a result of great pressure from labor and workers' groups, Hoeven's rider was dropped, and the silica rule was sent to the Office of Management and Budget for its final review within the ninety-day period provided by EO 12866, a Clinton-era executive order that requires that significant regulations be submitted to OIRA for review. At long last, on March 25, 2016, the Department of Labor announced the final silica rule, more than eighty years after Labor Secretary Frances Perkins identified silica dust as a deadly hazard.[179] Industry groups immediately filed petitions in federal court to block the rule, calling for still more studies, while giving lip service to safety. The National Association of Manufacturers issued the following statement: "Manufacturers are and have always been committed to safe workplaces, and we take pride in continuing to find ways to improve the work environment, but this unnecessary regulation is not the solution."[180]

Still more regulatory quicksand. In 1992, OSHA, under the Clinton Department of Labor, embarked on rule making for ergonomic workplace standards. Ergonomics is simply the science of preventing injuries to workers by designing the job to the person, and not the person to the job. When a workplace is designed without considering how it will physically affect the worker, the consequential results are injured workers, lost productivity, and skyrocketing workers' compensation and health care costs. In 1995, OSHA estimated that work-related musculoskeletal disorders resulted in more than seven hundred thousand lost work days. Despite evidence that an ergonomic standard is good for the business bottom line, a proposed rule was not announced until 1995, and a final rule was delayed until November 14, 2000, quite literally in the last hours of the Clinton presidency. Still, the long-awaited ergonomic standard went into effect on January 16, 2001, only four days before George W. Bush was sworn in as president. Flush with electoral victory, courtesy of the Supreme Court, the U.S. Chamber of Commerce wasted no time in demanding its just deserts.[181] In what was rightly billed as a resounding defeat for workers and labor, and a victory for the Chamber and its allies, Congress used the arcane Congressional Review Act (CRA) for the first time since it was enacted in 1996 and voted on March 6 and 7, 2001, to repeal the ergonomics standard. Ten years of rule making and scores of hearings and scientific testimony were wiped out. President Bush signed the disapproval resolution on March 20, 2001, and without irony declared: "The safety and health of our Nation's workforce is a priority for my administration. Together we will pursue a comprehensive approach to ergonomics that addresses the concerns surrounding the ergonomics rule repealed today. We will work with the Congress, the business community, and our Nation's workers to address this important issue."[182]

Since then, OSHA can only encourage employers to voluntarily implement effective ergonomic programs, and has developed voluntary guidelines. Enforcement of ergonomic hazards are under the General Duty Clause. Yet without an ergonomic standard to protect workers, more than 33 percent of all reported work injuries and illnesses continue to be attributable to MSDs.[183]

In 2002, and in response to a 1999 recommendation from a joint committee of labor, management, and public representatives, OSHA initiated rule making to update and strengthen the construction safety standards for cranes and derricks.[184] In 2004, and after negotiated rule making involving these stakeholders, a proposed rule was issued. However, a final rule was not proposed until October 2008 and, predictably, only after a series of deadly crane accidents occurred that spring in rapid succession, which claimed the lives of twelve workers. Still there was more delay, and the final rule was not issued until August 2010.[185] In the eight years of rule making, 176 workers died in crane accidents that could have been prevented if new standards had been in place.[186] Regrettably, this story did not end with the final rule. After the final rule was issued, OSHA gave the industry four years to comply, and required that it certify its compliance to OSHA. The required certification included whether a crane operator could safely operate a crane at a construction site. However, outside groups raised objections. And on September 26, 2014, OSHA announced that it would delay the final rule until November 10, 2017.[187] Not surprisingly, crane operators continue to die and get injured in construction site accidents. In May 31, 2015, a crane carrying a massive heating and air conditioning unit snapped under the weight of four tons and crashed down thirty stories onto midtown Manhattan, injuring ten persons.[188]

Beryllium is a gray metal, stronger than steel and lighter than aluminum. Its unique physical properties make it an essential material in the construction, aerospace, telecommunications, computer, medical, defense, and nuclear industries. It is also a known workplace hazard. When inhaled, beryllium can cause debilitating lung diseases, including cancer. OSHA has estimated that thirty-five thousand workers are exposed to beryllium at their work and in benign spaces like dental offices, where exposure occurs from the alloy in bridges, crowns, and dental plates.[189]

In August 2015, OSHA announced that it was proposing a new rule to update the permissible exposure limit to beryllium, fifteen years after the Department of Energy (DOE) finalized a level of 0.2 micrograms per cubic meter of air for workers exposed to beryllium at DOE facilities, and for DOE contractors. The current standard was established in 1948 by the Atomic Energy Commission and is ten times higher than the proposed, safer standard. OSHA claims that the new proposed standard can save one hundred lives and prevent fifty serious illnesses.[190] While welcome news, the proposed standard is a long way from a final rule, a process that can take years to complete.

Unbelievably, outside of the grain industry, there is no OSHA standard to prevent explosions from the accumulation and ignition of combustible dust. There are only OSHA guidelines.[191] Virtually any substance that becomes airborne through processing can turn into a deadly fireball. This includes flour to bake bread, and sugar to make candy. But combustible dust also commonly comes from metals, wood, and chemicals.

The history of the combustible-dust standard began, like many rules, from worker deaths. However, before October 21, 2009, when OSHA finally issued the Advanced Notice of Proposal Rulemaking that initiated rule making for combustible dust, the danger of this phenomenon was well known. In November 2006, the U.S. Chemical Safety Board (CSB) studied the hazard of combustible dust and recommended that OSHA issue a standard.[192] The George W. Bush administration ignored the CSB's recommendations, and more workers died in preventable workplace disasters. On April 30, 2008, in the aftermath of the Imperial Sugar disaster, the House of Representatives passed legislation mandating that OSHA issue a combustible-dust rule. The legislation was never passed by the Senate.[193] With Obama in the White House, rule making began again in October 2009, only to disappear from OSHA's regulatory agenda after the 2010 midterm elections.[194] The next required step in federal rule making is to convene a Small Business Regulatory Enforcement Fairness Act (SBREFA) panel. Initially scheduled for 2010, the SBREFA panel was postponed through October 2016.[195] Apparently cowed by the "complexity" of a combustible-dust standard in the last year of the Obama administration, OSHA kicked the regulatory can down the road, giving the U.S. Chamber of Commerce another victory.[196] While this delay in the OSHA combustible-dust standard dragged on, eight more workers died in preventable explosions in 2010 and 2011.[197]

In a final example of the dysfunction in rule making caused by regulatory delay, on July 8, 2016, a group of employer representatives, including the National Association of Manufacturers, filed suit in a Texas federal court challenging OSHA's new standard limiting employer safety incentive programs and allowing the agency to post online injury and illness data. In an all-too-familiar hyperbolic charge, counsel for the National Association of Manufacturers described the OSHA standard as "putting a target on nearly every manufacturer in this country." The purpose of the standard was to encourage employees to report a workplace injury or illness and to rein in incentive and drug-testing programs that discourage workers from reporting.[198]

These are just a few examples of the powerful opposition to any rules that will protect workers from known hazards, seemingly without regard to overwhelming scientific and economic evidence. Yet the opposition goes beyond substantive rules. On July 29, 2015, OSHA issued a Notice of Proposed Rulemaking that clarifies an employer's continuing commonsense obligation to make and maintain a record of each recordable work injury and illness for the most recent five-year period.[199] Known as the *Volks* rule, it clarified that employers have an ongoing obligation to report a recordable illness or injury for the five-year retention period. The purposes of recordkeeping are obvious. For one, it allows all stakeholders—employers, employees, OSHA, and researchers—to have records that will allow them to identify hazards at the workplace that are the most serious and that have caused injuries in the past, so these hazards can be eliminated in the future. For employees—assuming that they look into an employer's safety record before taking a job—the records may provide them with crucial information about the risks of working for a particular employer. Whatever the purpose, the clarification from OSHA was in response to a lawsuit challenging OSHA's authority to cite an employer for a recordkeeping violation more than six months after the obligation to report arose, usually the date of the accident. Weighing in on the lawsuit with an amicus curiae brief was the National Federation of Independent Business, another advocacy group with links to the Koch brothers, which received millions in funding from Karl Rove's political advocacy group, Crossroads GPS. In 2012, the D.C. Circuit Court of Appeals in, *AKM LLC v. Secretary of Labor (Volks)*, held that the OSH Act's six-month statute of limitations for issuing violations prevented OSHA from issuing a recordkeeping citation more than six months after the date the obligation to record arose. Accordingly, if an employer does not report a recordable illness or injury, OSHA's handful of investigators have six months to find the needle in the haystack, or be forever barred from issuing a citation to a noncompliant employer. The ongoing obligation is especially important given OSHA's limited resources to inspect work sites and, instead, its reliance on an honor system among employers.

The proposed rule does not add any new costs or burdens to employers, nor does it affect small businesses. The clarifying rule only covers large employers in the most dangerous industries, and clarifies the ongoing obligation to maintain accurate records, an obligation that has existed without challenge for the past 45 years. The rule was not controversial when it passed, and OSHA received only 27 comments during the comment period, including comments from employers supporting the rule.[200]

On December 19, 2016, OSHA issued a final *Volks* rule clarifying an employer's reporting obligation as continuing. However, on February 21, 2017, Congress, again using the CRA, introduced a joint resolution of disapproval of the *Volks* rule, seeking its repeal. The repeal of this sensible rule requiring accurate and timely reporting allows unsafe working conditions to go unreported and unremedied, and puts American workers in even greater harm's way.

In the early days of the Trump administration, and in an unprecedented attack on the regulatory authority of federal agencies, Congress used the CRA to repeal many Obama-era regulations in an effort to achieve, in the words of White House chief strategist Stephen K. Bannon, the "deconstruction of the administrative state."[201] Also repealed at this time using the CRA was the Fair Pay and Safe Workplaces rule, which was issued to implement an Obama executive order requiring government contractors receiving federal tax dollars to disclose in the procurement process violations of labor laws, including violations of the OSH Act.[202] The fact that an employer seeking a federal contract might be a recidivist would have been factor in considering whether to reward that employer with millions of dollars of taxpayer money. For the U.S. Chamber and its allies in Congress, such transparency was known as the "blacklisting rule," and down it went in the wake of Trump's election.[203]

In addition to congressional obstacles to regulations, the federal courts have proved fertile ground for more delay and regulatory rejection. Agency regulatory decisions are almost always made with one eye to the courts. As a result, regulators often propose rules so as to draw the least amount of fire from groups opposed to a particular regulation or decision and the inevitable court challenge. In addition, rules must be carefully crafted to meet existing jurisprudence.[204] But in today's hypercharged politicized environment, almost every rule, no matter how small, comes with an almost-certain risk of judicial challenge. And although judicial review generally favors an agency in interpreting a regulation, so long as it is reasonable and not arbitrary, and "can enable agencies to vindicate broad-based interests by interpreting legislation in ways most consonant with public interests," the specter of looming judicial challenges affects the regulatory process at every level, both substantively and procedurally.[205] Indeed, in March 2015, the Supreme Court decided in *Perez v. Mortgage Bankers Association* that an agency's interpretation of a regulation need not go through public notice and comment rule making even if it is a change from a prior interpretation. While concurring in the judgment, Justices Scalia and Thomas planted the seeds for future legal battles over judicial deference to an agency's decision, arguing that agencies can interpret to their heart's delight, but the federal courts should decide whether the interpretation is correct.[206] Although agencies may have won the day in *Perez*, it seemed clear, at least until Justice Scalia's death, that the Supreme Court would revisit the issue of deference in a case that is properly ginned up for decision—that is, in a case where controversy is created by dubious or artificial means. Nonetheless, for employer groups the line will be a long one of aspiring litigants, fronted by businesses claiming that the agencies have run amok and backed by the usual antigovernment suspects and their legions of lawyers, all promising more mischief, delay, and uncertainty to needed regulations that protect everything from our air and water and the food we eat, to the health and safety of Americans at work.

Finally, added to the landscape of regulatory dysfunction, is the U.S. Office of Information and Regulatory Affairs, a government office that few Americans have ever heard about, but one that can exercise great power in the regulatory world. OIRA is an arm of the White House and was established to provide overview of proposed agency rules to ensure that they comply with cost-benefit analyses, and, as important, to ensure that the rules are in line with a particular administration's regulatory goals and policies. OIRA can endorse or refuse to support an agency's rules and regulatory agenda. As a result, and depending on who occupies the White House, OIRA exerts great influence on the regulatory landscape by enabling a president in effect to veto an agency's regulations. Moreover, and because "only so many regulatory decisions can command the White House's sustained attention," delay is inevitable. Public Citizen reported that "chronic delays have become a persistent and unfortunate feature of the OIRA review process" even under President Obama.[207] Executive Order 12866, signed by President Clinton, and reaffirmed by President Obama, directs OIRA to spend no more than ninety days reviewing a proposed agency rule, and allows for a single thirty-day extension. Despite this order, Public Citizen reported that during the Obama administration between ninety and one hundred rules were under review each week at OIRA, and more than a quarter of them were delayed beyond the speedy review process established in the executive order (including the silica rules), adding up to more than five thousand days of delay.[208]

The regulatory quicksand that Dodd-Frank has unceremoniously sunk into is an ominous case study for ordinary American workers. If Wall Street can scurry and escape the white-hot light of public anger from millions of Americans who lost their homes, jobs, and health in the 2008 financial crash, what chances do a foundry worker inhaling silica, a crane operator at a construction site, or a worker in a sugar refinery have of OSHA's making final rules to protect them at their workplace? Who can effectively advocate for them in the halls of Congress against the unlimited coffers, lobbyists, and lawyers of the regulated industries?

Given all these problems with the ability of regulations to be part of a system that can help protect Americans at work, what does real regulatory reform look like? And is real regulatory reform even possible? The later question assumes that "real reform" encompasses the ability of regulations to benefit most Americans and to make their lives safer and more secure, including their work lives. But for the moment, and as earlier discussed, actual regulatory reform looks a lot different from "real reform," as defined above.

In 2015 there were hundreds of Senate and House bills that proposed in one way or another "reform" through deregulation, or less regulation. One such bill, the Regulation Sensibility through Oversight Restoration Resolution of 2015, or RESTORE, was introduced on March 20, 2015, by Senator Mike Rounds, a South Dakota Republican, and cosponsored by eleven Senate Republicans and one Democrat. According to GovTrack. us, through 2016 there were 12,057 "similar bills" to RESTORE, including an identical bill introduced in the House.[209] Almost all were introduced by Republicans, and all have Orwellian sounding names as Searching for and Cutting Regulations That Are Unnecessarily Burdensome Act of 2015 (SCRUB Act, introduced by Senator Orrin Hatch of Utah); Regulations from the Executive in Need of Scrutiny Act of 2015 (REINS Act, introduced by Senator Rand Paul of Kentucky); Stop the EPA Act of 2015 (introduced by Representative Sam Graves of Missouri); Defending Rivers from Overreaching Policies Act of 2015 (introduced by Senator Jeff Flake of Arizona); and, the most ironic of all, the Duplicative Elimination Act of 2015 (introduced by Representative Charles Dent of Pennsylvania). Earlier Congresses since 2010 have introduced and reintroduced similar legislation. All the hundreds of pieces of faux regulatory "reform" legislation have only one goal: to make it more difficult to implement needed regulations that protect, among other things, the air, water, food, consumer products, cars, and workplaces of Americans from known dangers.

RESTORE is a good example of how this type of legislation undermines needed regulations. RESTORE proposes to create in Congress a Joint Select Committee on Regulatory Reform for the purpose of reviewing and approving all rules that have an annual "significant" effect on the economy of over $50 million before an agency can promulgate a final rule.[210] This sounds reasonable, except that the low $50 million threshold is half the current $100 million threshold, which was established in 1978 and has never been updated. Beyond the unreasonably low dollar threshold, Congress has neither the scientific technical expertise, nor time and ability, to review the thousands of proposed rules that are critical to public health and safety. Which is precisely the point. RESTORE is designed to prevent any proposed regulations from seeing the light of day. RESTORE's sponsor, Senator Mike Rounds, elected in 2014, recycled well-worn antigovernment talking points in announcing this bill, declaring that it was time to end "regulation without representation" and that his bill would lift the heavy hand of government from the collective spirit of hardworking Americans.[211] Rounds repeated previously debunked statistics, prominently including the allegation that regulations are a "hidden tax" costing more than $1.88 trillion per year, or $15,000 per American family.[212] The latter figure Rounds sourced from the Competitive Enterprise Institute, another advocacy group funded by the Koch family foundations, as well as other notable conservative billionaires and fossil fuel companies.[213] Significantly, figures repeated by Rounds ignored the economic benefits of regulations, a part of the economic equation that deregulators often ignore. In 2016, the Office of Management and Budget reported that from 2005 to 2015, the cost-benefit of over a one hundred significant regulations had a combined value of the benefits that far exceeded their combined costs, from a high of $672 billion in benefits to a high of $85 billion in costs.[214]

Without a doubt, regulatory reform is a vexing problem. Given the political hostility to regulations from Republicans, as well as some Democrats, nothing short of a complete end to political contributions would be a good start. But at this juncture in America, ending political bribery is a fantasy. Where does this leave us? Is there a work-around to this regulatory dysfunction that, in particular, will help protect American workers?

RESTORE and other phony "reform" bills introduced in greater numbers since the 2010 midterm elections have one goal: to delay or shut down entirely the regulatory process behind the ruse of more analysis, especially by diverting attention to the costs of regulation while ignoring the economic benefits. A good start at real reform would be to streamline the regulatory process so that much-needed rules can be quickly finalized and enforced. One way to streamline the process is to adopt and implement, as regulatory rules, those existing voluntary practices that have been accepted in a particular industry as a safe, or best, practice to alleviate a known safety or health hazard. But then the antiregulatory chorus will chime: Why not just let the "market" enforce voluntary practices by rewarding the safest companies, without the heavy hand of government? The answer is, for the same reason that the "market" doesn't regulate how we drive our cars. A person's decision to drive while drunk affects more than the impaired driver. Companies, like drivers, need to know the rules of the road for their industry. In cases where the industry at large has created a broad consensus as to a particular best practice, why must there be a deliberately drawn-out process before the practice can be codified as a final and enforceable regulation? The regulatory process must be freed from the grip of the so-called free-marketers who have never seen a regulation that they like. Regulations on lead, asbestos, seat belts, fuel tanks, ozone, clean air, and cigarettes, to name just a very few, have saved lives and helped Americans remain healthier without raining down economic ruin. The rule of reason, which once prevailed, must be restored.

Once final regulations have been issued, there must be a broad and nonpartisan commitment to enforcement. This is critical in the real world of workplace safety where there are not enough inspectors to visit every work site. Frankly, Americans have to demand that violators be appropriately identified and punished. Doing this would require the enhancement of criminal prosecutions, as previously discussed, as well as vigorous whistleblower protections, so workers are not afraid to report bad and dangerous work practices.

In the end, and in the current political crisis in America, especially in Washington, real regulatory reform seems about as remote a likelihood as there can be. In fact, the antiregulatory and antigovernment zealots have nearly achieved their goals with the election of Donald Trump. On the first day of the Trump administration, the White House issued a memorandum instructing federal agencies to freeze the effective dates of any new regulations.[215] Trump followed the freeze with an executive order requiring that federal agencies eliminate two existing regulations for every new regulation implemented, so any new rule must have a zero net cost, and without regard to whether there are any economic benefits from a rule.[216] Known as the "two-for-one" executive order, it is part of Trump's campaign promise to eliminate 75 percent of regulatory rules.[217]

In the current political environment, there is virtually no political space for real regulatory reform. Instead, lip service to workers from the "Workers' President" is deemed good enough, while American workers continue to get sick, get injured, and die at unacceptable rates. As a consequence, we must pivot our focus from inside the Beltway.

Change outside the Beltway: Tip O'Neill famously said, "All politics is local." This elementary statement simply means that a politician's electoral success is directly tied to his or her ability to understand and respond to the personal issues affecting constituents. Whether this means fixing the potholes on streets, or bringing home the political bacon of jobs for a taxpayer-funded roads project, a politician must be responsive to the needs of those who vote.

How can this principle be applied to the issue of workplace safety? The simple answer is, real reforms affecting workplace safety have to occur at the state and local levels, and outside the Beltway. For reasons exemplified by the repeal of the Fair Pay and Safe Workplaces federal rule, advocates must continue to train their focus and resources away from Washington and our elected representatives in Congress and the White House. This is not to say that the powerful and moneyed interests arrayed against real reform won't follow the fight. But city officials, or even congressional representatives, are more likely to respond to an organized grassroots effort with a personal and direct connection to their ward or district than a grander, national effort that is more dispersed in its impact and easier to defeat. In the worker-safety world, a proposed OSHA regulation that seeks to address a known safety hazard that affects workers all over the United States is easy to pick off and block in Congress. But fights over reforms and laws across fifty states, and literally thousands of cities and towns, are another matter. There are not enough lobbyists and lawyers to attend every city council meeting or town hall event. Real reform will have to occur on the local level. One village, one town, one city at a time. Successful examples of this strategy are scattered across the map of America.

Every year, state and local governments publicly finance millions of jobs throughout the United States by entering into contracts that cost hundreds of billions of dollars. Many of these jobs pay low wages, expose workers to unsafe workplaces, and result in labor law and employment violations, including misclassification and wage and hour violations. In response, more and more state and local governments are enacting "responsible contracting ordinances" (RCOs), also called responsible employer ordinances and responsible bidder ordinances. These RCOs are codified best practices for publicly financed contracts that, among other things, require compliance with labor and employment laws.[218] A survey by the Associated Builders and Contractors Inc., which opposes RCOs, reported in 2012 that dozens of local governments in twelve states have adopted some version of an RCO, including the cities of Los Angeles, Chicago, Boston, and Philadelphia, as well as the State of New York.[219]

In 2014, Maryland[220] and Minnesota[221] joined the list, passing their own RCOs. In Minnesota, its RCO applies to publicly financed projects in excess of $50,000. Among other requirements for getting a contract paid by taxpayer money, a contractor or subcontractor must certify that it has been in compliance with federal and state wage laws for the prior three years and has not violated any laws regarding the classification or misclassification of workers.[222]

In addition to focusing on compliance with existing labor and employment laws, more than 140 local governments require the payment of higher wages and better benefits through "living wage" laws, which are defined using federal poverty guidelines. Houston and San Francisco, for example, require that public contractors provide health benefits to their employees, or pay into a fund to offset the public costs of uninsured workers.[223]

The common denominator in all RCOs is that these best practices ensure that public contracting and the expenditure of public monies benefit both taxpayers and workers. By receiving higher wages and benefits, and by working in safer conditions, workers are less likely to need the public dole to support them in the form of wage assistance, public health benefits for the uninsured, and unemployment benefits due to lost time from work injuries.[224]

When it comes to worker safety, RCOs must include provisions that bar, or debar, any contractor from public contracts who has a history of workplace injuries and safety violations. The City of Los Angeles has a responsible-contractor policy that requires bidders to complete a detailed questionnaire on their history of compliance with environmental, labor, and safety laws. The Santa Clara Valley Transportation Authority has a pre-qualification process, and only firms that have good safety records, among other metrics, are allowed to bid on public contracts.[225] Some laws also require that contractors provide safety training on a public project.

Although I have urged an emphasis on state and local legislation, the White House and Congress remain a priority, as well, especially when the White House and the Department of Labor are occupied and staffed by like-minded persons. In this regard, one of President Obama's most praised and criticized strategies was the use of executive orders, especially in the waning years of his administration. Regarding workplace health and safety, on July 31, 2014, Obama issued the Fair Pay and Safe Workplaces Executive Order requiring that employers with federal contracts in excess of $500,000 must have complied with labor laws, including the OSH Act, for the previous three-year period.[226] As expected, the executive order met stiff resistance from the U.S. Chamber of Commerce and the National Association of Manufacturers, to name just a few. The GOP-controlled Congress joined the fight by refusing to authorize funds to create a new agency to ensure compliance with the order.[227] Still, according to the Labor Department, the final rules made "sure that agencies have the information they need to determine which contractors are providing their workers with basic protections."[228] Under the Trump administration and a GOP-controlled Congress, a strategy that relies on Washington is not a viable one.

While RCOs and other similar contracting laws apply only to public contracts, their effect is more widespread. Most obviously, any contractor or subcontractor who wants to do business in the billion-dollar public-sector world must also be a responsible contractor in the private-sector world. The opportunity to make money has a powerful influence on a business's behavior, including its commitment to a safe workplace. As taxpayers, Americans must demand that contractors receiving public monies certify compliance with health and safety laws. If it is important that our lattes and diamonds be fair-trade certified and sustainable, shouldn't we demand the same for our public jobs? Santa Clara and the thousands of burgs around the United States are where the battle for workplace safety reforms must be fought. All politics is local.

7

ARE THERE REALLY ANY ACCIDENTS?

Because going to work shouldn't be a grave mistake.
UNITED SUPPORT AND MEMORIAL FOR
WORKPLACE FATALITIES

Let us review what we now know. We know that in the roulette of worker safety that exists today in America, thirteen workers die each day in acute work-related incidents, many more are seriously injured, and hundreds more will eventually succumb to occupational diseases. We know that when these deaths and injuries happen, American families are left to fend for themselves against insurance companies and their armies of adjusters, doctors, and lawyers, as they navigate the medical and compensation systems and, in the end, more often than not, are left impoverished and emotionally spent. We know that powerful political forces and their courtiers have unparalleled access to the levers of power in America, and that they purposefully act in ways that are knowingly harmful to workers. What do we do with all this knowledge?

Circle back to the central questions posed at the beginning of the book, and, knowing what we now know, try to answer them. Are thirteen deaths each day an acceptable number in order to power the American economy? And have we done everything we can to protect ourselves, our loved ones, and our neighbors? The answer to both questions, knowing what I now know, is quite clearly no. We can do much better; we can do much more.

Social justice requires, in addition to proximity and knowledge gained, a change in the dominant narrative. If there is one constant that has remained embedded in our vernacular when discussing workplace health and safety, it is the notion that workers are killed and injured in "accidents." Merriam-Webster defines an accident as "a sudden event that is not planned or intended and that causes damage and injury." Other definitions include "an event that occurs by chance," which is a lack of design. *Black's Law Dictionary* similarly defines an accident, and uses the phrase "unforeseen and unexpected," as well as "not deliberate and not inevitable." OSHA avoids using the word "accident" altogether and instead refers to its cousin, an "incident." Fatality and Catastrophe Summaries, also called "OSHA 170 Forms," are used by OSHA after an inspection of a reported fatality or catastrophe to describe the incident and its causes. Using the word "accident" promotes a false narrative that workplace deaths and injuries could not have been prevented. Yet nothing can be further from the truth.

All the stories in this book, and the thousands more that can be easily found on websites and blogs like the *Weekly Toll*, hosted by the United Support and Memorial for Workplace Fatalities (USMWF), or Public Citizen's *Workplace Health & Safety Digest*, are not about deaths and injuries caused by accidents.

OSHA publishes a weekly fatality report. At the time of this writing, the most current data was a fatality report for the week ending December 19, 2015, which reported nineteen workplace deaths. For the fiscal year to date, which began on October 4, 2015, there were 227 deaths,[1] or more than 20 deaths per week. And again, these are only the reported deaths.

I don't want to end this book with more numbing statistics and numbers. Instead, the following is a list of the names of the workers who died on the job from the last week of fiscal year 2015.[2] OSHA's website reports only fatality cases under its jurisdiction, which excludes public employees like police, firefighters, and miners, who are covered by a different statute, the Federal Coal Mine Health and Safety Act. Moreover, the list includes only those work-related deaths that were reported *and* investigated. Feel free to view OSHA's website to get a current list.

TABLE 7.1 Regional Federal and State Fatality / Catastrophe Weekly Report

Date of incident	Company	Victim	Cause of death
10/3/15	TNT International Ltd., Newark, NJ	Vitaliy Popovych	Worker struck and killed by forklift
10/2/15	All City Construction, Newark, NJ	John Burke	Worker killed in fall from roof
10/2/15	Bigge Crane and Rigging Co., Webster, CA	Giuseppe Santuccio	Worker killed in vehicle collision
10/2/15	Middle Tennessee Recycling LLC, Hermitage, TN	Billy Earl Lowe	Worker fatally crushed by machine used to move rolls of wire
10/1/15	PDQ Welding & Machining, Ontario, CA	Saul (no last name provided)	Worker fatally crushed by machinery
10/1/15	Sheehan Pipeline Construction Co., Tulsa, OK	Brandon Jermaine Mitchell	Worker struck and killed by falling tree first day on the job
10/1/15	Tecta America SE Region, Orange City, FL	Robert Heyman	Worker in aerial lift fatally crushed between lift basket and roof overhang
10/1/15	Troy Builders Construction and Remodeling, Los Angeles, CA	Salvador Ortiz Gonzalez	Worker killed in fall from roof
9/30/15	Argix Direct Inc., Jamesburg, NJ	N/A	Worker struck and killed by tractor trailer cab
9/30/15	Tenaris Coiled Tubes LLC, Houston, TX	David Magana	Worker fatally crushed between spools of tubing and forklift
9/29/15	Goodwill of North Georgia, Conyers, GA	Pamela Harmon	Worker fatally shot by former employee
9/29/15	Nickels and Dimes Inc., Las Vegas, NV	Chuck Wyman	Worker electrocuted while servicing arcade machine
9/28/15	A Rooter Man of Pittsburgh, Butler, PA	Jacob Casher	Worker killed in trench collapse
9/28/15	East Coast Contracting Inc., Wilmington, NC	Manuel Melendez and James Williams	Workers electrocuted when excavator boom contacted power line

(*Continued*)

TABLE 7.1 (*Continued*)

Date of incident	Company	Victim	Cause of death
9/28/15	J&R Transport Inc., Baltimore, MD	Charles Sibert	Worker struck and killed by trusses that fell off delivery truck
9/28/15	Stillwater Plumbing, Odenton, MD	Kenneth Oyola	Worker killed in fall from ladder

Like the on-the-job fatalities recounted in this book, each of these workers suffered a horrific death and left behind family, friends, and co-workers. Pick any random week from OSHA's weekly summary and you'll find the same grisly toll.

The workplace injuries and deaths described in this book are not "unforeseen and unexpected" and certainly were not "inevitable." Hanna Phillips's amputation was not an accident. According to OSHA, by September 2015, forty-six workers in Arkansas suffered amputations at work in fiscal year 2015.[3] At Tyson Foods, there was on average more than one amputation per month in the nine-month period of January 1, 2015, through September 30, 2015. One worker lost both hands at Tyson's Saint Joseph, Missouri, plant.[4] It was foreseen, expected, and inevitable that Paul King would get electrocuted because he was not properly trained and was not supplied with inexpensive thermal gloves. It was not by chance that Scott Shaw drowned. Rather, it was because his barge employer did not provide him with a life preserver and had jerry-rigged the barges together. Eufracia Barrera's repetitive-motion injuries, and the hundreds of injuries suffered by her coworkers, were predictable and could have been avoided if her employer had implemented procedures to reduce or mitigate the epidemic of MSDs among its workers. The National Institute for Occupational Safety and Health investigated deaths and injuries in the energy industry, particularly in the Bakken oil fields in North Dakota and Montana. There, on average, a worker dies every six weeks. So it was no surprise that Ray Gonzalez died, along with Leonard Moore Jr., or that workers were blown off the oil platform in the BP/Transocean Deepwater Horizon disaster. Finally, in the blazing summer of 2014, and at a time when there are still many politicians who deny climate change, eighteen workers died from heatstroke and related causes, while 2,630 suffered from heat illness at work.[5]

In addition to not accurately describing how the death and injury occurred, using the word "accident" pushes the blame and ownership for the death and injury onto the worker, where it does not belong. If there is a single takeaway from this book, it is this: there needs to be a real shift in how workplace deaths and injuries are described, and we must start by holding accountable those persons responsible for the deaths and injuries that are not accidents, and stop blaming the worker.

But still we blame, or partially blame, the worker. In June 2015, the Texas Supreme Court held, in a twist of logic that would have made Lewis Carroll smile like a Cheshire cat, that an employer is shielded from liability for a work-related death or injury to an employee if the employee was aware of the defect or hazard that led to the death or injury.[6] The court wrote: "An employer generally does not have a duty to warn or protect its employees from unreasonably dangerous premises conditions that are open and obvious or known to the employee." Remember the train conductor who long ago fell because of a rickety handrail? By this measure, it was his fault. In other words, according to the Texas Supreme Court, it's the fault of the injured worker because he or she was aware of the hazard. It's not the fault of the employer who caused or allowed the hazard and exposed the employee to it. Up is down. Down is up. In a damning editorial by the *Houston Chronicle* in July 2015, the paper decried a culture that ignores worker safety amid the "incessant whining by our elected officials about overregulation," and noted that Texas employers are not required to carry workers' compensation insurance, while "more construction workers die on the job in the Lone Star State than in any other state." The *Chronicle* called for more to be done to encourage safe workplaces, because in Texas, "lives and livelihoods depend on it."[7]

Changing the narrative will help change our mind-set from the notion that workers die and get injured in accidents. Yet this is no easy task. It requires a more heightened awareness than currently exists: that all workers, to one degree or another, are daily exposed to preventable workplace hazards.

Despite the political treachery of our elected officials, and the regulatory deconstruction that leaves workers at risk, changing the narrative is still possible. Diana Florence's story is one example of this possibility. Florence is a veteran prosecutor of twenty-one years in the Manhattan District Attorney's office. She is a high energy New Yorker, born on Long Island, and she talks passionately about her work and providing justice to victims of crimes and their families. Florence started out in 1995 doing the usual tour of duty, beginning with prosecuting street crimes, muggings, robberies and drug-related offences. After 9/11, Florence worked on cases involving fraud perpetrated by people posing as relatives of victims in order to obtain their benefits. In 2007, she moved over to construction and labor racketeering fraud and eventually became the attorney in charge of the Construction Fraud Task Force, a team of federal, state and city lawyers and investigators with jurisdiction across all five boroughs.

Florence's first big case with the Task Force was the successful prosecution of a materials testing company, Testwell Laboratories, that falsified thousands of tests assessing the strength of concrete used in major public projects, including the Second Avenue subway, One World Trade Center, the new Yankee Stadium, and other sites around New York. Testwell's president was sentenced to significant jail time, and a successor testing company, American Standard Testing, was indicted and prosecuted for doing the same thing.

The spotlight that Florence and her team put on the highly lucrative and well connected New York construction industry made a real difference in the materials testing industry. After a few prosecutions, the industry cleaned up its act and loopholes were closed in the decades old Building Code.

On April 6, 2015, Florence set her sights on a different kind of construction crime—homicide. Carlos Moncayo, a 22-year-old Ecuadoran immigrant, was working at a construction site in the fashionable Meatpacking District, home to many of Manhattan's best restaurants and exclusive nightclubs. Moncayo was crushed to death in an unsecured trench collapse at a retail development site for the trendy Restoration Hardware chain. Trench collapses at construction sites are rarely survivable because a single cubic yard of soil can weigh up to 3,000 pounds. But trench collapses are preventable if a contractor follows well established safety standards by shoring up any trench that is more than five feet deep. From 2012 through 2016, OSHA reported a marked increase in fatalities from trench collapses. By mid-November 2016, there were 23 deaths reported from trench collapses, far outstripping any previous year. In almost every case, the collapse would be chalked up as another accident. At best, OSHA might come in to investigate and impose a fine.

But Florence and her team of investigators did not treat Carlos Moncayo's death as an accident. When the death was reported, one of Florence's top investigators immediately went to the site and recognized that there was more to Moncayo's death than just another construction site accident. Florence was contacted by the investigator, and they all met at the precinct house that same day and began an investigation into a possible crime. Florence's investigators learned that a site supervisor had repeatedly warned the contractor and subcontractor that the trench was not properly supported, but still workers like Moncayo were sent down into the hole. The supervisor eventually quit over his unheeded safety warnings and later became a critical witness for Florence. In the end, Florence charged the foreman, Wilmer Cueva, with criminally negligent homicide for creating a danger and disregarding the warnings, as well as manslaughter and reckless endangerment. Cueva was acquitted of manslaughter but convicted of negligent homicide and reckless endangerment. He was later sentenced to one to three years in prison. After a hung jury in the trial of another supervisor, Alfonso Prestia, Cueva eventually pled guilty to criminally negligent homicide, but he received no jail time.

Florence acknowledged that Cueva was not a criminal mastermind but, she said, "someone has to be accountable for the preventable death of a twenty-two-year-old." During the sentencing phase, numerous contractors sent letters to the court supporting the conviction. These contractors spend a lot of time and money on safety, Florence explained, and they understood that getting crushed in a thirteen-foot-deep unsupported trench was not an accident.

Florence and her task force do more than prosecute. They have started an outreach program to workers and safety advocates in the construction industry to teach them to report safety hazards, explaining that construction deaths can be more than just an accident; they can be a crime. The task force is also deeply involved in drafting legislation in the city and state to make workers safer. In the end, Florence says, "what we hope to accomplish are not just prosecutions, but a change of mind set." For Florence and her task force, the narrative has changed. As a result, construction workers in New York are safer than before.[8]

A new awareness begins with proximity and with using the pronoun "we." As in, we are all at risk for death and injury in our workplaces. We are Paul King. We are Hanna Phillips. We are Ray Gonzalez. We are Carlos Moncayo. Workplace health and safety must become a shared responsibility, because we are all at risk. It can start at home, and in our communities, and at our workplaces. The stories collected in this book are but the tip of the proverbial iceberg. The stories themselves do not even represent a statistical sampling. Rather, they are meant to illustrate that every single day millions of Americans working in hundreds of job occupations are exposed to death and injury. Tragically there is an endless supply of stories about death and injury in the American workplace, enough to fill dozens of more books. And, sadly, all these deaths and injuries were preventable. There are no accidents.

Epilogue

Most good stories have an arc to their plotline, usually without any loose ends and unfinished business at the conclusion. Cliffhangers eventually run out of cliffs to dangle over. Happy endings are what most of us yearn for, in our books and in our lives, even if getting there was frightful and dangerous. All the better to exhale a sigh of relief at the end, certain as we turn the last page that our heroine is safe and sound and will live happily ever after as we read silently to ourselves, *The End*.

Not so for American workers when it comes to their health and safety at work. And, as a consequence, there is no *The End* in this book. There is literally no end to the story of American workers dying and getting injured at work. There is only a pause, and time to catch our breath before reading about the next worker killed on the job or forever maimed. No happy endings there. Sure, there are some exceptions. There are some good, if not truly happy moments for the workers and their families in this book. Hannah Phillips has a beautiful boy. Kathy Rodriguez, Tammy Miser, and many others have channeled their anger and grief toward helping survivors like them. But there is really no end to the daily toll of death and destruction affecting American workers.

Writing *Dying to Work*, I had to make a mostly arbitrary decision about when to stop seeking out and interviewing injured workers and surviving family members, in order to finish the book. Horribly, there is no shortage of material. Among other sources, I subscribe to Listservs, occupational safety blogs, and news services. Even after deciding that I was done writing and sending the book off to be published, I still read the daily e-mails and links to the stories that have no end, and struggle with being pulled back in to write just a little bit more. Often, I save the new material in my research files, just in case.

Although, for me, this particular book is finished, American workers remain at high risk of injury and death. So, in the end, it remains my hope that *Dying to Work* is a beginning. A beginning to a dialogue. A beginning to awareness. A beginning to action. A beginning to meaningful change. And, most hopefully, a beginning to an end.

NOTES

Introduction

1. Unless otherwise specifically noted, when I refer to workplace safety, or a safe workplace, I am referring as well to a healthy workplace, one free from injury and death due to exposure.

2. Sometimes I refer to "American workers." This is not meant to refer only to workers born or naturalized in the United States. Rather, it is shorthand for all persons working in the United States, including the approximately 8.1 million unauthorized immigrants. Jens Manuel Krogstad, Jeffery S. Passel, and D'Vera Cohen, "5 Facts about Illegal Immigration in the U.S.," the website of the Pew Research Center, September 20, 2016, http://www.pewresearch.org/fact-tank/2016/09/20/5-facts-about-illegal-immigration-in-the-u-s/.

3. David Ropeik, "How Risky Is Flying?," *NOVA* online, October 17, 2006, http://www.pbs.org/wgbh/nova/space/how-risky-is-flying.html.

4. Brad Plumer, "Eight Facts about Terrorism in the United States," *Wonkblog*, *Washington Post*, April 16, 2013, https://www.washingtonpost.com/news/wonk/wp/2013/04/16/eight-facts-about-terrorism-in-the-united-states/.

5. John Mueller and Mark G. Stewart, "Terror, Security, and Money: Balancing the Risks, Benefits, and Costs of Homeland Security," paper presented at the annual convention of the Midwest Political Science Association, Chicago, April 1, 2011, http://politicalscience.osu.edu/faculty/jmueller//MID11TSM.PDF.

6. "Census of Fatal Occupational Injuries Summary, 2015 (CFOI)," website of the Bureau of Labor Statistics, U.S. Department of Labor, December 16, 2016, http://www.bls.gov/news.release/cfoi.nr0_.htm.

7. Nearly twice as many Americans have been killed by non-jihadist white terrorists as by jihadi-inspired murders. Scott Shane, "Homegrown Extremists Tied to Deadlier Toll Than Jihadists in U.S. since 9/11," *New York Times*, June 24, 2015, http://www.ny times.com/2015/06/25/us/tally-of-attacks-in-us-challenges-perceptions-of-top-terror-threat.html?_r=0.

8. "Census of Fatal Occupational Injuries Summary, 2015."

9. *Death on the Job: The Toll of Neglect; National and State-by-State Profile of Worker Safety and Health in the United States*, 25th ed., report by the AFL-CIO, April 2016, http://www.aflcio.org/content/download/174867/4158803/1647_DOTJ2016.pdf.

10. "Nonfatal Occupational Injuries and Illnesses Requiring Days Away from Work, 2015," news release from the Bureau of Labor Statistics, U.S. Department of Labor, November 10, 2016, http://www.bls.gov/news.release/pdf/osh2.pdf.

11. *Death on the Job*, 25th ed.

12. The notion of becoming proximate was eloquently discussed in a talk I heard from Bryan Stevenson, the great public-interest lawyer who has fought and challenged poverty and racial discrimination in the criminal justice system.

1. America Goes to Work

1. Charles Hirschman and Elizabeth Mogford, "Immigration and the American Industrial Revolution from 1880 to 1920," *Social Science Research* 38, no. 4 (December 1, 2009): 897–920, https://www.ncbi.nlm.nih.gov/pmc/articles/PMC2760060/.

2. Sukkoo Kim, "Immigration, Industrial Revolution, and Urban Growth in the United States, 1820–1920: Factor Endowments, Technology, and Geography," working paper, National Bureau of Economic Research, January 2007, http://www.nber.org/papers/w12900.

3. Mark Aldrich, "History of Workplace Safety in the United States, 1880–1970," encyclopedia entry, EH.net, Economic History Association, August 14, 2001, https://www.eh.net/encyclopedia/history-of-workplace-safety-in-the-united-states-1880-1970-2/.

4. Mark Aldrich, *Safety First: Technology, Labor, and Business in the Building of American Worker Safety, 1870–1939* (Baltimore: Johns Hopkins University Press, 1997).

5. Ibid.

6. The information in this and the following paragraphs about low wages and dangerous conditions for immigrant workers is taken from David von Drehle, *Triangle: The Fire That Changed America* (New York: Grove, 2003).

7. *New York Times*, December 14, 1909.

8. Von Drehle, *Triangle*, 66.

9. Ibid., 77.

10. Ibid., 86.

11. Information in this and the following paragraph is from Richard A. Greenwald, "'More Than a Strike': Ethnicity, Labor Relations, and the Origins of the Protocol of Peace in the New York Ladies' Garment Industry," *Business and Economic History* 27, no. 2 (Winter 1998): 318–29.

12. Von Drehle, *Triangle*, 47.

2. The Torch That Lighted Up the Industrial Scene

1. Information in this chapter about the Triangle Shirtwaist Factory and fire is from David von Drehle, *Triangle: The Fire That Changed America* (New York: Grove, 2003).

2. Judson MacLaury, "Government Regulation of Workers' Safety and Health, 1877–1917," U.S. Department of Labor, https://www.dol.gov/general/aboutdol/history/mono-regsafeintrotoc.

3. Ibid.; Von Drehle, *Triangle*, 195.

4. MacLaury, "Government Regulation of Workers' Safety."

5. Frances Perkins, *People at Work* (New York: John Day, 1934), 50.

6. The information in this and the following paragraphs, about workers' compensation and other Progressive Era reforms, is from MacLaury, "Government Regulation of Workers' Safety."

3. Keeping Americans Safe at Work

1. Mark Aldrich, *Safety First: Technology, Labor, and Business in the Building of American Worker Safety, 1870–1939* (Baltimore: Johns Hopkins University Press, 1997), 6.

2. Judson MacLaury, "Government Regulation of Workers' Safety and Health, 1877–1917," U.S. Department of Labor, https://www.dol.gov/general/aboutdol/history/mono-regsafe introtoc.

3. Paul Underwood Kellogg, ed., *The Pittsburgh Survey: Findings in Six Volumes* (Philadelphia: Press of Wm. F. Fell Co., 1914), https://archive.org/stream/pittsburghsurvey05kell uoft/pittsburghsurvey05kelluoft_djvu.txt.

4. "The Russell Sage Foundation and the Pittsburgh Survey," *Women Working, 1800–1930* (online resource), Harvard University Library Open Collections Program, http://ocp.hul.harvard.edu/ww/rsf.html.

5. Crystal Eastman, *Work-Accidents and the Law*, vol. 2 of Kellogg, *Pittsburgh Survey*, 14–15, https://www.russellsage.org/sites/default/files/Eastman%26Kellog_Work%20Acci dents_0.pdf.

6. Jonathan Grossman, "The Origin of the U.S. Department of Labor," *Monthly Labor Review*, March 1973, https://www.dol.gov/oasam/programs/history/dolorigabridge.htm.

7. Judson MacLaury, "The Job Safety Law of 1970: Its Passage Was Perilous," *Monthly Labor Review*, March 1981, http://www.bls.gov/opub/mlr/1981/03/art2full.pdf.

8. Aldrich, *Safety First*, xvii.

9. Ibid., 69.

10. Ibid., 211–12.

11. Ibid., 257.

12. Chris Hamby, "Black Lung Surges Back in Coal Country," Center for Public Integrity, July 8, 2012, updated May 19, 2014, https://www.publicintegrity.org/2012/07/08/9293/black-lung-surges-back-coal-country.

13. Grossman, "Origin of the U.S. Department of Labor."

14. MacLaury, "Job Safety Law of 1970."

15. Gregg LaBar, "Seven Decades of Safety: Good Times Take Their Toll," *EHS Today*, October 1, 2008, http://ehstoday.com/safety/management/good_times_toll.

16. OSHA covers private-sector employers and employees.

17. MacLaury, "Job Safety Law of 1970."

18. Ibid.

19. Ibid.

20. 29 U.S. Code § 654(a)1, "Duties of Employers and Employees," https://www.law.cornell.edu/uscode/text/29/654?qt-us_code_temp_noupdates=0#qt-us_code_temp_noupdates.

21. David Burnham, "Worker Safety Agency, under Fire, Has Little Impact but Big Potential," *New York Times*, December 20, 1976, http://www.nytimes.com/1976/12/20/archives/work-safety-agency-under-fire-has-little-impact-but-big-potential.html.

22. Owen Ullmann, "Two Senators Team-Up to Limit OSHA," Associated Press, December 28, 1979.

23. Stephen Labaton, "OSHA Leaves Worker Safety in Hands of Industry," *New York Times*, April 25, 2007, http://www.nytimes.com/2007/04/25/washington/25osha.html.

24. Anthony DePalma, "Air Masks at Issue in Claims of 9/11 Illnesses," *New York Times*, June 5, 2006, http://www.nytimes.com/2006/06/05/nyregion/05masks.html.

25. "Protection for Employer Self-Audits," U.S. Chamber of Commerce, August 4, 2010, https://www.uschamber.com/protection-employer-self-audits.

26. OpenSecrets.org, Center for Responsive Politics, https://www.opensecrets.org/lobby/clientsum.php?id=D000019798&year=2008.

27. OpenSecrets.org, Center for Responsive Politics, https://www.opensecrets.org/out sidespending/detail.php?cycle=2012&cmte=US+Chamber+of+Commerce.

28. J. Paul Leigh, "Economic Burden of Occupational Injury and Illness in the United States," *Milbank Quarterly* 89, no. 4 (December 2001): 728–72, doi:10.1111/j.1468-0009.2011.00648.x.

29. National Safety Council, *Journey to Safety Excellence: The Business Case for Investment in Safety, A Guide for Executives* (pamphlet), 2013, http://www.nsc.org/JSEWork placeDocuments/Journey-to-Safety-Excellence-Safety-Business-Case-Executives.pdf.

30. Thomas J. Donohue, "Regulatory Relief Is on the Way," *Above the Fold / In Your Corner* (blog), U.S. Chamber of Commerce website, February 27, 2017, https://www.uscham ber.com/above-the-fold/regulatory-relief-the-way.

4. Just the Facts

1. "Census of Fatal Occupational Injuries Summary, 2015," website of the Bureau of Labor Statistics, U.S. Department of Labor, December 16, 2016, http://www.bls.gov/news.re lease/cfoi.nr0.htm.

2. "Employer-Reported Workplace Injuries and Illnesses 2014," Bureau of Labor Statistics, October 29, 2015.

3. *Death on the Job: The Toll of Neglect; National and State-by-State Profile of Worker Safety and Health in the United States*, 24th ed., report by the AFL-CIO, April 2015, http://www.aflcio.org/content/download/154671/3868441/DOTJ2015Finalnobug.pdf.

4. "Census of Fatal Occupational Injuries Summary, 2015 (CFOI)."

5. *Death on the Job: The Toll of Neglect; National and State-by-State Profile of Worker Safety and Health in the United States*, 25th ed., report by the AFL-CIO, April 2016, pp. 2, 9, http://www.aflcio.org/content/download/174867/4158803/1647_DOTJ2016.pdf.

6. Ted Oberg and Trent Seibert, "OSHA in Houston: A Worker's Watchdog with No Teeth," *ABC13 Eyewitness News*, November 17, 2015, http://abc13.com/news/osha-a-workers-watchdog-with-no-teeth/1088852/.

7. Federal Annual Monitoring and Evaluation (FAME) Reports by State, United States Department of Labor, Occupational Safety and Health Administration, www.osha.gov/dcsp/osp/efame.

8. "FY 2014 Follow-Up Federal Annual Monitoring and Evaluation (FAME) Report, State of Indiana," U.S. Department of Labor, Occupational Health and Safety Administration, Region V, Indianapolis, https://www.osha.gov/dcsp/osp/efame/2014/in_report_2014.pdf.

9. *Death on the Job*, 25th ed., 20.

10. Ibid., 21.

11. Ibid., 24–25.

12. In January 2017, using the Congressional Review Act, Republicans in Congress introduced a disapproval resolution to repeal an Obama rule that clarified OSHA's ability to issue citations against companies for recordkeeping violations that are discovered more than six-months after the last violation.

13. *Hidden Tragedy: Underreporting of Workplace Injuries and Illness*, report by the Committee of Education and Labor, U.S. House of Representatives, June 2008, http://www.

bls.gov/iif/laborcommreport061908.pdf. See also Sara E. Wuellner and David K. Bonauto, "Exploring the Relationship between Employer Recordkeeping and Underreporting in the BLS Survey of Occupational Injuries and Illnesses," *American Journal of Industrial Medicine* 57, no. 10 (August 5, 2014): 1133–43, https://www.ncbi.nlm.nih.gov/pmc/articles/PMC4260691/.

14. *Hidden Tragedy*, 2. See also John W. Ruser, "Examining Evidence of Whether BLS Undercounts Workplace Injuries and Illnesses," *Monthly Labor Review*, August 2008, http://www.bls.gov/opub/mlr/2008/08/art2full.pdf.

15. William J. Wintroski, "Examining the Completeness of Occupied Injury and Illness Data: An Update on Current Research," *Monthly Labor Review*, June 2014, http://www.bls.gov/opub/mlr/2014/article/examining-the-completeness-of-occupational-injury-and-illness-data-an-update-on-current-research-1.htm.

16. *Workplace Safety and Health: Enhancing OSHA's Records Audits Process Could Improve the Accuracy of Worker Injury and Illness Data*, U.S. Government Accountability Office report to Congressional Requesters, October 2009, http://www.gao.gov/new.items/d1010.pdf.

17. J. Paul Leigh, James P. Marcin, and Ted R. Miller, "An Estimate of the U.S. Government's Undercount of Nonfatal Occupational Injuries," *Journal of Occupational and Environmental Medicine* 46, no. 1 (January 2004), http://journals.lww.com/joem/Abstract/2004/01000/An_Estimate_of_the_U_S__Government_s_Undercount_of.4.aspx?trendmd-shared=0.

18. Ibid.

19. David Michaels, *Year One of OSHA's Severe Injury Reporting Program: An Impact Evaluation*, March 17, 2016, https://www.osha.gov/injuryreport/2015.pdf.

20. *Hidden Tragedy*, note 12.

21. Baruch Fellner, "Statement of the U.S. Chamber of Commerce on Underreporting of Workplaces Injuries and Illnesses to the House Committee on Education and Labor," June 19, 2008, 2–3, http://www.gibsondunn.com/publications/Documents/Fellner-OSHATestimony061908.pdf.

22. Ibid., 7.

23. "OSHA's Recordkeeping Rule: Changes to Reporting Requirements," U.S. Department of Labor, Occupational Health and Safety Administration, https://www.osha.gov/recordkeeping2014/.

24. "Key Labor, Employment, and Immigration Regulatory Initiative in the Obama Administration," U.S. Chamber of Commerce, Labor, Immigration, and Employee Benefits Division, May 13, 2015, https://www.uschamber.com/sites/default/files/5132015_reg_issues_update_for_lrc.pdf.

25. James McPherson, "OSHA: North Dakota Oil Worker Killed in Crane Accident," Associated Press, June 21, 2016, http://bigstory.ap.org/article/80ebb00924134d409ca9536d887fbbfb/osha-north-dakota-oil-worker-killed-crane-accident.

26. Celeste Monforton, "Fatal Worker Injury That Killed Kenneth Schultz Was Preventable, Cal-OSHA Cites Labor Ready," *Pump Handle*, Science Blogs, June 14, 2016, http://scienceblogs.com/thepumphandle/2016/06/14.

27. "Worker Killed in Carnival Ride Accident in New York," Associated Press, September 2, 2012, http://www.foxnews.com/us/2012/09/02/worker-killed-in-carnival-ride-accident-in-new-york.html.

28. Robbyn Mitchell, "Adventure Island Employee Dies after Being Struck by Lightning," *Tampa Bay Times*, September 10, 2011, http://www.tampabay.com/news/publicsafety/adventure-island-employee-dies-after-being-struck-by-lightning/1190902.

29. Dave Dreeszen, "Tyson Faces $104,200 in Fines for Dakota City Worker Death," *Sioux City Journal*, August 27, 2012, http://siouxcityjournal.com/business/local/tyson-faces-in-fines-for-dakota-city-worker-death/article_7aa065f9-43d2-57e0-82ad-46fdbf1c7edf.html.

30. "Boy Dies after Getting Stuck in Shrimping Vessel Winch," Associated Press, August 12, 2012, http://www.wtvy.com/home/headlines/Boy-Dies-After-Getting-Stuck-in-Shrimping-Vessel-Winch-165926206.html.

31. *Final Investigation Report: Chevron Richmond Refinery Pipe Rupture and Fire*, U.S. Chemical Safety and Hazard Investigation Board, report no. 2012-03-I-CA, January 2015, http://www.csb.gov/assets/1/19/Chevron_Final_Investigation_Report_2015-01-28.pdf.

32. Barb Ickes, "Mount Carroll in Shock over Grain Bin Deaths," *Quad-City Times*, July 29, 2010, http://qctimes.com/news/local/mount-carroll-in-shock-over-grain-bin-deaths/article_0ea98abe-9b8c-11df-8ed0-001cc4c002e0.html.

33. "Water, Rest, Shade: The Work Can't Get Done without Them," U.S. Department of Labor, Occupational Health and Safety Administration, https://www.osha.gov/SLTC/heatillness/index.html.s.

34. David Michaels (assistant secretary for occupational safety and health, U.S. Department of Labor), in a letter to Sidney Wolfe (Public Citizen's Health Research Group), June 7, 2012, http://www.citizen.org/documents/denial-of-heat-stress-petition.pdf.

5. Stories

1. Marc Levinson, *The Great A&P and the Struggle for Small Business in America* (New York: Farrar, Straus and Giroux, 2011).

2. Paul B. Ellickson, "The Evolution of the Supermarket Industry: From A&P to Walmart," unpublished paper, March 15, 2015, http://paulellickson.com/SMEvolution.pdf.

3. Ibid.

4. "Supermarket Facts," Food Marketing Institute, http://www.fmi.org/research-resources/supermarket-facts.

5. "Guidelines for Retail Grocery Stores: Ergonomics for the Prevention of Musculoskeletal Disorders," U.S. Department of Labor, Occupational Safety and Health Administration, 2004, https://www.osha.gov/ergonomics/guidelines/retailgrocery/retailgrocery.html.

6. "The American Hotel and Lodging Association: 1910–Present," AHLA website, https://www.ahla.com/american-hotel-lodging-association-1910-present.

7. Ibid.

8. "Lodging Industry Trends 2015," AHLA website, https://www.ahla.com/sites/default/files/Lodging_Industry_Trends_2015.pdf.

9. "Housekeeping Can Be Dangerous Work" (fact sheet), UNITE HERE website, http://unitehere.org/wp-content/uploads/HousekeepingDangerous.pdf.

10. Susan Buchanan et al., "Occupational Injury Disparities in the U.S. Hotel Industry," *American Journal of Industrial Medicine* 53, no. 2 (February 2010), doi:10.1002/ajim.20724.

11. "Housekeeping Can Be Dangerous Work."

12. Richard E. Fairfax (deputy assistant secretary, U.S. Department of Labor), letter to Thomas J. Pritzker (CHB-CEO, Hyatt Corporation), April 25, 2012, http://algfoiafiles.com/images/2/25/OSHA_letter_to_Hyatt_04.25.12.pdf.

13. U.S. Department of Labor, Occupational Safety and Health Administration, Regulations, Standards 29 CFR, Standard Number 1910.269, https://www.osha.gov/pls/oshaweb/owadisp.show_document?p_table=STANDARDS&p_id=9868.

14. "Workplace Electrical Injury and Fatality Statistics, 2003–2010," Electrical Safety Foundation International, http://www.esfi.org/resource/workplace-electrical-injury-and-fatality-statistics-2003-2010-280.

15. "Occupational Employment and Wages May 2015, 47-2111 Electricians," Occupational Employment Statistics, U.S. Department of Labor, Occupational Safety and Health Administration, http://www.bls.gov/oes/current/oes472111.htm.

16. U.S. Department of Labor, Occupational Safety and Health Administration, Regulations, Standards 29 CFR, standard number 1910.399, https://www.osha.gov/pls/oshaweb/owadisp.show_document?p_table=STANDARDS&p_id=9976.

17. "Standard for Electrical Safety in the Workplace," Standard 70E, National Fire Protection Association, http://www.nfpa.org/codes-and-standards/all-codes-and-standards/list-of-codes-and-standards?mode=code&code=70E.

18. "Timeline of Coal in the United States," American Coal Foundation, 2005, www.teachcoal.org/lessonplans/pdf/coal_timeline.pdf.

19. "Historical Coal Mine Disasters, 1900–2015" (table), Centers for Disease Control and Prevention, www.cdc.gov/niosh/mining/UserFiles/statistics/15g05aac.svg.

20. "Injury Trends in Mining," U.S. Department of Labor, Mine Safety and Health Administration, www.msha.gov/MSHAINFO/FactSheets/MSHAFCT2.HTM.

21. Ibid.

22. "Injuries, Illnesses, and Fatalities: Fact Sheet, Coal Mining," U.S. Department of Labor, Bureau of Labor Statistics, April 2010, www.bls.gov/iif/oshwc/osh/os/osar0012.htm.

23. "Respirable Dust Rule: A Historic Step Forward in the Effort to End Black Lung Disease," U.S. Department of Labor, Mine Safety and Health Administration, https://www.msha.gov/news-media/special-initiatives/2016/09/28/respirable-dust-rule-historic-step-forward-effort-end.

24. 30 U.S.C. Sec. 922.

25. U.S. Department of Labor, Office of Workers' Compensation Programs, Division of Coal Mine Workers' Compensation, www.dol.gov/owcp/dcmwc/regs/compliance/blbene.htm.

26. "Poverty Guidelines," U.S. Department of Health and Human Services, Office of the Assistant Secretary for Planning and Evaluation, January 25, 2016, https://aspe.hhs.gov/poverty-guidelines.

27. Chris Hamby, "Black Lung Surges Back in Coal Country," Center for Public Integrity, July 8, 2012 (updated May 19, 2014), https://www.publicintegrity.org/2012/07/08/9293/black-lung-surges-back-coal-country.

28. Ibid.

29. "Lowering Miners' Exposure to Respirable Coal Mine Dust, Including Continuous Personal Dust Monitors," Office of the Federal Register, May 1, 2015, https://www.federalregister.gov/documents/2014/05/01/2014-09084/lowering-miners-exposure-to-respirable-coal-mine-dust-including-continuous-personal-dust-monitors.

30. U.S Energy Information Administration, www.eia.gov/tools/faqs/faq.cfm?id=23&+=10.

31. Rakteem Katakey, "U.S. Ousts Russia as Top World Oil, Gas Producer in BP Data," *Bloomberg News*, June 10, 2015, http://www.bloomberg.com/news/articles/2015-06-10/u-s-ousts-russia-as-world-s-top-oil-gas-producer-in-bp-report.

32. U.S. Energy Information Administration, www.eia.gov/todayinenergy/.

33. *Oil and Natural Gas Stimulate American Economic and Job Growth*, Vendor Survey Findings Report, 2014, American Petroleum Industry, http://www.api.org/~/media/files/policy/jobs/oil-gas-stimulate-jobs-economic-growth/api-state-vendor-survey-findings-report.pdf.

34. "Oil and Gas Extraction," U.S. Department of Labor, Occupational Safety and Health Administration, Safety and Health Topics, https://www.osha.gov/SLTC/oilgaswelldrilling/.

35. Krystal L. Mason, Kyla D. Retzer, Ryan Hill, and Jennifer M. Lincoln, "Occupational Fatalities during the Oil and Gas Boom: United States, 2003–2013," Morbidity and Mortality Weekly Report, Centers for Disease Control and Prevention, May 29, 2015, https://www.cdc.gov/mmwr/preview/mmwrhtml/mm6420a4.htm.

36. Robert J. Harrison et al., "Sudden Deaths among Oil and Gas Extraction Workers Resulting from Oxygen Deficiency and Inhalation of Hydrocarbon Gases and Vapors: United States, January 2010–March 2015," Morbidity and Mortality Weekly Report, Centers for

Disease Control and Prevention, January 15, 2016, http://www.cdc.gov/mmwr/volumes/65/wr/mm6501a2.htm.

37. Jennifer Gollan, "In North Dakota's Bakken Oil Boom, There Will Be Blood," *Reveal*, Center for Investigative Reporting, June 13, 2015, https://www.revealnews.org/article/in-north-dakotas-bakken-oil-boom-there-will-be-blood/.

38. Ibid.

39. "Revisions to the 2014 Census of Fatal Occupational Injuries (CFOI)," website of the Bureau of Labor Statistics, U.S. Department of Labor, last modified April 21, 2016, http://www.bls.gov/iif/cfoi_revised14.htm.

40. Alexandra Berzon, "Oil Deaths Rise as Bakken Boom Fades," *Wall Street Journal*, March 12, 2015, http://www.wsj.com/articles/oil-deaths-rise-as-bakken-boom-fades-1426187062.

41. Gollan, "In North Dakota's Bakken Oil Boom."

42. Berzon, "Oil Deaths Rise."

43. National Park Service, "Mississippi River Facts," https://www.nps.gov/miss/riverfacts.htm.

44. "Navigation," U.S. Army Corps of Engineers, www.usace.army.mil/Missions/Civil-Works/Navigation.

45. "Prelude FLNG" page, Shell Global website, www.shell.com/about-us/major-projects/prelude-flng.html.

46. "Dredging and Dredged Material Management," U.S. Army Corps of Engineers, July 31, 2015, http://www.publications.usace.army.mil/Portals/76/Publications/Engineer Manuals/EM_1110-2-5025.pdf.

47. "Corps of Engineers Launches New Water Safety Campaign, 'Life Jackets Worn, Nobody Mourns,'" U.S. Army Corps of Engineers, March 3, 2016, http://www.lrn.usace.army.mil/Media/News-Stories/Article/685864/corps-of-engineers-launches-new-water-safety-campaign-life-jackets-worn-nobody/.

48. *Deck Barge Safety*, OSHA 3358 01N 2009, U.S. Department of Labor, Occupational Safety and Health Administration, https://www.osha.gov/Publications/3358deck-barge-safety.pdf.

49. Ibid.

50. Logging e-tool, U.S. Department of Labor, Occupational Safety and Health Administration, https://www.osha.gov/SLTC/etools/logging/.

51. "Logging Safety," Centers for Disease Control and Prevention, National Institute for Occupational Safety and Health, www.cdc.gov/niosh/topics/logging/.

52. "Revisions to the 2014 Census of Fatal Occupational Injuries (CFOI)."

53. "FRA Safety Alerts," National Timber Harvesting and Transportation Safety Foundation, Logging and Transportation Safety, http://www.loggingsafety.com/safety_alerts.

54. "Logging Workers," Occupational Outlook Handbook, U.S. Department of Labor, Bureau of Labor Statistics, http://www.bls.gov/ooh/farming-fishing-and-forestry/logging-workers.htm.

55. *Sugar Dust Explosion and Fire*, U.S. Chemical Safety and Hazard Investigation Board, Investigation Report No. 2008-05-I-GA, September 2009, http://www.csb.gov/UserFiles/file/Imperial%20Sugar%20Report%20Final%20updated.pdf.

56. Ibid., 3.

57. Ibid., 20.

58. Ibid., 21.

59. Ibid., 22.

60. *Combustible Dust Hazard Study*, U.S. Chemical Safety and Hazard Investigation Board, Investigation Report No. 2006-H-1, November 2006, http://www.csb.gov/assets/1/19/dust_final_report_website_11-17-06.pdf.

61. "Worker Protection against Combustible Dust Explosions and Fires Act of 2008," H.R. 5522, https://www.congress.gov/bill/110th-congress/house-bill/5522/text.

62. "Rulemaking Watch: OSHA's Combustible Dust Rulemaking," U.S. Chemical Safety Board, www.csb.gov/mobile/mw/dust/regwatch/.

63. Ibid.

64. Rafael Moure-Eraso, "The Danger of Combustible Dust," *New York Times*, August 22, 2014, http://www.nytimes.com/2014/08/23/opinion/the-danger-of-combustible-dust.html?_r=0.

65. Warehouse Workers for Justice, fact sheet, www.warehouseworker.org/industry.html.

66. Jason Struna et al., "Unsafe and Unfair: Laboratory Conditions in the Warehouse Industry," *Policy Matters* 5, no. 2 (Summer 2012), http://www.policymatters.ucr.edu/pmatters-vol5-2-warehouse.pdf.

67. "Poultry Processing," U.S. Department of Labor, Occupational Safety and Health Administration, Safety and Health Topics, https://www.osha.gov/SLTC/poultryprocessing/index.html.

68. National Association of Manufacturers, www.nam.org/Newsroom/Facts-About-Manufacturing/.

69. "Manufacturing: NAICS 31–33," U.S. Department of Labor, Bureau of Labor Statistics, Industries at a Glance, www.bls.gov/iag/tgs/iag31–33.htm.

70. "Chapter 1: Basics of Machine Safeguarding," U.S. Department of Labor, Occupational Safety and Health Administration, https://www.osha.gov/Publications/Mach_Safe Guard/chapt1.html.

71. "Manufacturing: NAICS 31–33."

72. Luis Felipe Martínez, "Can You Hear Me Now? Occupational Hearing Loss, 2004–2010," *Monthly Labor Review*, July 2012, https://www.bls.gov/opub/mlr/2012/07/art4full.pdf.

73. "Noise and Hearing Loss Prevention," Centers for Disease Control and Prevention, National Institute for Occupational Safety and Health, https://www.cdc.gov/niosh/topics/noise/stats.html.

74. "Grain Handling," U.S. Department of Labor, Occupational Safety and Health Administration, Safety and Health topics, https://www.osha.gov/SLTC/grainhandling/.

75. Ibid.

76. Ibid.

77. Steve Riedel and Bill Field, "2010 Summary of Grain Entrapments in the United States" (online paper), February 9, 2011, p. 1, https://extension.entm.purdue.edu/grainlab/content/pdf/2010GrainEntrapments.pdf.

78. Ibid., 2–3.

79. Ibid., 5.

80. Salah Issa, Yuan-Hsin Cheng, and Bill Field, "2014 Summary of U.S. Agricultural Confined Space-Related Injuries and Fatalities" (online paper), December 8, 2011, https://extension.entm.purdue.edu/grainsafety/pdf/Space_Confined_Summary_2014.pdf.

81. Florence Nightingale, *Notes on Nursing: What It Is, and What It Is Not* (New York: D. Appleton and Co., 1860).

82. "Nonfatal Occupational Injuries and Illnesses Requiring Days Away from Work, 2015," Economic News Release, November 10, 2016, Bureau of Labor Statistics, https://www.bls.gov/news.release/osh2.nr0.htm.

83. Keith Wrightson and Taylor Lincoln, *Uplifting an Industry? State-Based Safe Patient Handling Laws Have Yielded Improvements but Are Not Adequately Protecting Health Care Workers*, report for Public Citizen, June 24, 2015, http://www.citizen.org/documents/part-three-state-health-care-worker-safety-laws-uplifting-industry.pdf.

84. Ibid.

85. Roni Jacobson, "Epidemic of Violence against Health Care Workers Plagues Hospitals," *Scientific American*, December 31, 2014, https://www.scientificamerican.com/article/epidemic-of-violence-against-health-care-workers-plagues-hospitals/.

86. "OSHA Takes Steps to Prevent Violence in Healthcare Workplaces," Safety.BLR, Workplace Safety News, December 14, 2015, http://safety.blr.com/workplace-safety-news/equipment-and-process-safety/healthcare-safety/OSHA-takes-steps-to-prevent-violence-in-healthcare/.

87. *Guidelines for Preventing Workplace Violence for Healthcare and Social Service Workers*, U.S. Department of Labor, Occupational Safety and Health Administration, OSHA 3148-04R 2015, https://www.osha.gov/Publications/osha3148.pdf.

88. Dianna Hunt, Sherry Jacobson, and Holly K. Hacker, "Nurses: Hospital's Ebola Response Puts Workers, Patients at Risk," *Dallas Morning News*, October 15, 2014, http://www.dallasnews.com/news/news/2014/10/15/nurses-hospitals-ebola-response-put-workers-patients-at-risk.

89. T. Grimmond and L. Good, "EXPO-S.T.O.P.-2012: Year Two of a National Survey of Sharps Injuries and Mucocutaneous Blood Exposures among Healthcare Workers in U.S. Hospitals," *Journal of the Association of Occupational Professionals in Healthcare*, Spring 2015, http://www.aohp.org/aohp/Portals/0/MembersOnlyDocuments/survey%20reslut/EXPOSTOP2012SurveyResult.pdf.

90. Yasser Sakr et al., "The Impact of Hospital and ICU Organizational Factors on Outcome in Critically Ill Patients: Results from the Extended Prevalence of Infection in Intensive Care Study," *Critical Care Medicine* 43, no. 3 (March 2015): 519–26.

91. Patricia W. Stone, Sean P. Clarke, Jeannie Cimiotti, and Rosaly Correa-de-Araujo, "Nurses' Working Conditions: Implications for Infections Disease," *Emerging Infectious Diseases* 10, no. 11 (November 2014), https://www.ncbi.nlm.nih.gov/pmc/articles/PMC3328993/.

92. Charles R. Figley, ed., *Compassion Fatigue: Coping with Secondary Traumatic Stress Disorder in Those Who Treat the Traumatized* (New York: Routledge, 1995).

93. Brenda Sabo, "Reflecting on the Concept of Compassion Fatigue," *Online Journal of Issues in Nursing*, January 31, 2011, http://www.nursingworld.org/MainMenuCategories/ANAMarketplace/ANAPeriodicals/OJIN/TableofContents/Vol-16-2011/No1-Jan-2011/Concept-of-Compassion-Fatigue.html.

94. "Burj Khalifa" entry, Skyscraper Center online resource, http://skyscrapercenter.com/building/burj-khalifa/3.

95. Amy Frearson, "Top Nine Tallest Skyscrapers Completing in 2016," *dezeen* online, January 2, 2016, http://www.dezeen.com/2016/01/02/top-9-tallest-skyscrapers-completing-in-2016-roundup/.

96. Michael McCann, *Deaths and Injuries Involving Elevators and Escalators*, Center for Construction Research and Training, September 2013, p. 2, http://www.cpwr.com/sites/default/files/publications/elevator_escalator_BLSapproved_0.pdf.

97. Ibid., 4.

98. Lorena Mongelli, "Elevator Haters," *New York Post*, December 28, 2011, http://nypost.com/2011/12/28/elevator-haters/.

6. What Can We Do?

1. "Koch and George Mason University," *DeSmog* blog, http://www.desmogblog.com/koch-and-george-mason-university.

2. Eric Lichtblau, "Cato Institute and Koch Brothers Reach Agreement," *Caucus* blog, *New York Times*, June 25, 2012, http://thecaucus.blogs.nytimes.com/2012/06/25/cato-institute-and-koch-brothers-reach-agreement/.

3. Board of Directors, Cato Institute, https://www.cato.org/board-of-directors.

4. David R. Henderson, *Are We Ailing from Too Much Deregulation?*, Cato Policy Report, November/December 2008, https://www.cato.org/policy-report/novemberdecember-2008/are-we-ailing-too-much-deregulation.

5. "About Mercatus" web page, Mercatus Center, George Mason University, https://www.mercatus.org/about.

6. "Mercatus Board of Directors" web page, Mercatus Center, George Mason University, https://www.mercatus.org/board.

7. "History and Timeline" web page, Mercatus Center, George Mason University, https://www.mercatus.org/content/history-and-timeline.

8. "Mercatus Board of Directors."

9. Tess VandenDolder, "It's Not Just Politics: The Koch Brothers Give Big Bucks to Influence Higher Education Too," DCInno website, September 12, 2014, http://dcinno.streetwise.co/2014/09/12/its-not-just-politics-the-koch-brothers-give-big-bucks-to-influence-higher-education-too/.

10. "About *Regulation* Magazine," Cato Institute, http://www.cato.org/regulation/about.

11. Thomas J. Kniesner and John D. Leeth, "Abolishing OSHA," *Regulation* 4 (1995): 46–56, https://object.cato.org/sites/cato.org/files/serials/files/regulation/1995/10/v18n4-5.pdf.

12. "Commonly Used Statistics," U.S. Department of Labor, Occupational Safety and Health Administration, https://www.osha.gov/oshstats/commonstats.html.

13. John Leeth and Nathan Hale, "Evaluating OSHA's Effectiveness and Suggestions for Reform," Mercatus Center, April 23, 2013, https://www.mercatus.org/publication/evaluating-oshas-effectiveness-and-suggestions-reform.

14. Ibid.

15. "Commonly Used Statistics."

16. Leeth and Hale, "Evaluating OSHA's Effectiveness."

17. David Michaels, "OSHA Does Not Kill Jobs; It Helps Prevent Jobs from Killing Workers," *American Journal of Industrial Medicine* 55 (2012): 961–63, https://www.osha.gov/as/opa/michaels_commentary.html.

18. Kniesner and Leeth, "Abolishing OSHA."

19. John Leeth, "OSHA's Role in Promoting Occupational Safety and Health," Mercatus Center working paper, November 13, 2012, https://www.mercatus.org/publication/oshas-role-promoting-occupational-safety-and-health.

20. Ibid.

21. 29 U.S. Code § 651, "Congressional Statement of Findings and Declaration of Purpose and Policy," https://www.law.cornell.edu/uscode/text/29/651.

22. Henderson, *Are We Ailing?*

23. Ibid.

24. Leeth, "OSHA's Role in Promoting Occupational Safety."

25. Henderson, *Are We Ailing?*

26. Thomas A. Robinson, "New Study Points to Significant Underreporting of Injuries to Bureau of Labor Statistics," LexisNexis, August 29, 2014, https://www.lexisnexis.com/legalnewsroom/workers-compensation/b/recent-cases-news-trends-developments/archive/2014/08/29/new-study-points-to-significant-underreporting-of-injuries-to-bureau-of-labor-statistics.aspx.

27. Gabriel Brunswick, Michael A. Livermore, Richard L. Revesz, and Rachel A. Weise, Institute for Policy Integrity, New York University School of Law, "Comments on the Forthcoming Injury and Illness Prevention Program Rule (RIN 1218-AC48)," letter to David Michaels (U.S. assistant secretary of labor), August 7, 2013, http://policyintegrity.org/documents/Policy_Integrity_Letter_on_I2P2.pdf.

28. Samira J. Simone, "Mining Disaster Raises Questions about Effectiveness of Safety Laws," CNN, April 9, 2010, http://www.cnn.com/2010/US/04/09/west.virginia.mine.safety/.

29. Ben Casselman, "Closer Look at Union vs. Nonunion Workers' Wages," *Real Time Economics* blog, *Wall Street Journal*, December 17, 2012, http://blogs.wsj.com/economics/2012/12/17/closer-look-at-union-vs-nonunion-workers-wages/.

30. Scott Schneider, "Yes, Union Construction Really Is Safer," *Laborers' Health and Safety Fund of North America* 12, no. 5 (October 2015), http://www.lhsfna.org/index.cfm/lifelines/october-2015/yes-union-construction-really-is-safer/.

31. David Moberg, "Fatalities Higher at Non-Union Mines—Like Massey's Upper Big Branch," *In These Times*, April 9, 2010, http://inthesetimes.com/working/entry/5813/fatalities_higher_at_non-union_mineslike_masseys_upper_branch.

32. Daniel Malloy, "Are Union Mines Safer?," *Pittsburgh Post-Gazette*, April 18, 2010, http://www.post-gazette.com/local/region/2010/04/18/Are-union-mines-safer/stories/201004180265.

33. *Injustice on Our Plates*, report, Southern Poverty Law Center, November 7, 2010, https://www.splcenter.org/20101108/injustice-our-plates.

34. Jill Mislinski, "The Ratio of Part-Time Employed Remains High, but Improving," Advisor Perspectives website, November 7, 2016, http://www.advisorperspectives.com/dshort/updates/2016/11/07/the-ratio-of-part-time-employed-remains-high-but-improving.

35. "Union Members Summary," Economic News Release, Bureau of Labor Statistics, U.S. Department of Labor, January 28, 2016, http://www.bls.gov/news.release/union2.nr0.htm.

36. Leeth, "OSHA's Role in Promoting Occupational Safety."

37. Howard Berkes and Michael Grabell, "Injured Workers Suffer as 'Reforms' Limit Workers' Compensation Benefits," National Public Radio, March 4, 2015, http://www.npr.org/2015/03/04/390441655/injured-workers-suffer-as-reforms-limit-workers-compensation-benefits.

38. "Indiana Advantages," Indiana Economic Development Corporation, http://iedc.in.gov/indiana-advantages.

39. "Employment and Wages Online Annual Averages, 2015," web publication, Bureau of Labor Statistics, U.S. Department of Labor, last modified September 8, 2016, http://www.bls.gov/cew/cewbultn15.htm.

40. "National Census of Fatal Occupational Injuries in 2013," news release, Bureau of Labor Statistics, U.S. Department of Labor, September 11, 2014, http://www.bls.gov/news.release/archives/cfoi_09112014.pdf.

41. Matt Berger, "A Visual History of the U.S. Workforce, 1970 to 2012," *Marketplace*, Minnesota Public Radio, October 11, 2012, https://www.marketplace.org/2012/10/11/economy/visual-history-us-workforce-1970-2012.

42. Staffing Industry Statistics, American Staffing Association, https://americanstaffing.net/staffing-research-data/fact-sheets-analysis-staffing-industry-trends/staffing-industry-statistics/.

43. C.K. Smith et al., "Temporary Workers in Washington State," *American Journal of Industrial Medicine* 53, no. 2 (February 2010): 135–45, doi:10.1002/ajim.20728.

44. "Protecting Temporary Workers," U.S. Department of Labor, Occupational Safety and Health Administration, https://www.osha.gov/temp_workers/.

45. David Weil, *The Fissured Workplace: Why Work Became So Bad for So Many and What Can Be Done to Improve It* (Cambridge, MA: Harvard University Press, 2014), 9.

46. Protecting America's Workers Act, S. 1112, 114th Congress, https://www.congress.gov/bill/114th-congress/senate-bill/1112.

47. "Protecting America's Workers Act, S. 1112, 114th Congress," www.GovTrack.us, https://www.govtrack.us/congress/bills/114/s1112.

48. Senate hearing 110–895, "When a Worker Is Killed: Do OSHA Penalties Enhance Workplace Safety?," Committee on Health, Education, Labor, and Pensions, 110th Congress, second session, April 29, 2008, https://www.gpo.gov/fdsys/pkg/CHRG-110shrg42210/html/CHRG-110shrg42210.htm.

49. Statement by David Michaels (assistant secretary for occupational safety and health, U.S. Department of Labor), at "Protecting American's Workers Act: Modernizing OSHA Penalties," hearing before the Subcommittee on Workforce Protections, Committee on Education and Labor, U.S. House of Representatives, 111th Congress, second session, March 16, 2010, https://www.gpo.gov/fdsys/pkg/CHRG-111hhrg55302/pdf/CHRG-111hhrg55302.pdf.

50. 29 U.S. Code § 666, "Civil and Criminal Penalties," https://www.law.cornell.edu/uscode/text/29/666.

51. OSHA Instruction, Directive No. CPL 02-00-160, Field Operations Manual, August 2, 2016, https://www.osha.gov/OshDoc/Directive_pdf/CPL_02-00-160.pdf.

52. "2012. OSHA—Willful Violation of a Safety Standard Which Causes Death to an Employee," *U.S. Attorney's Manual*, Office of the United States Attorneys, U.S. Department of Justice, https://www.justice.gov/usam/criminal-resource-manual-2012-osha-willful-violation-safety-standard-which-causes-death.

53. OSHA Instruction, Directive No. CPL 02-00-160.

54. Ibid., 4–20.

55. 18 U.S. Code § 3571(c)(4), "Sentence of Fine," https://www.law.cornell.edu/uscode/text/18/3571.

56. OSHA Instruction, Directive No. CPL 02-00-160, pp. 3–3, 6–16.

57. Senate hearing 110–895, "When a Worker Is Killed," 16–17.

58. Martha T. McCluskey, Thomas O. McGarity, Sidney Shapiro, and Katherine Tracy, *OSHA's Discount on Danger: OSHA Should Revise Its Informal Settlement Policies to Maximize the Deterrent Value of Citations*, Center for Progressive Reform, June 2016, http://progressivereform.org/articles/OSHA_Discount_on_Danger_Report.pdf.

59. "Ohio Auto Parts Manufacturer Faces $3.42M in Fines after OSHA Finds Company Willfully Exposed Temporary Workers to Machine Hazards," OSHA Regional News Brief, Region 5, Occupational Safety and Health Administration, U.S. Department of Labor, June 29, 2016, https://www.osha.gov/pls/oshaweb/owadisp.show_document?p_table=NEWS_RELEASES&p_id=32736.

60. McClusky et al., *OSHA's Discount on Danger*, 16.

61. OSHA Instruction, Directive No. CPL 02-00-149, "Severe Violator Enforcement Program," June 18, 2010, https://www.osha.gov/pls/oshaweb/owadisp.show_document?p_table=DIRECTIVES&p_id=4503.

62. J. Paul Leigh, "Economic Burden of Occupational Injury and Illness in the United States," *Milbank Quarterly* 89, no. 4 (December 2011): 728–72, doi:10.1111/j.1468-0009.2011.00648.x.

63. Eric E. Hobbs, "2013 OSHA Developments and Review—What's New?," presentation at the Pulp and Paper Safety Association meeting, June 12, 2012, p. 23, http://www.ppsa.org/assets/2013Conference/Presentations/use-2013%20osha%20developments%20%20review%20whats%20new%20-%20eric%20hobbs.pdf.

64. Ibid., 29.

65. Ibid., 46.

66. Senate hearing 110–895, "When a Worker Is Killed," 23.

67. Statement by David Michaels, "Protecting America's Workers Act."

68. David Michaels (assistant secretary for occupational safety and health, U.S. Department of Labor), congressional testimony before the Committee on Health, Education, Labor

and Pensions, U.S. Senate, April 27, 2010, https://www.osha.gov/pls/oshaweb/owadisp. show_document?p_table=testimonies&p_id=1122.

69. Avi Meyerstein, "OSHA Gets Authorization to Boost Penalty Amounts," *National Law Review* online edition, December 13, 2015, http://www.natlawreview.com/article/osha-gets-authorization-to-boost-penalty-amounts.

70. "Research Finds that OSHA Citations, Penalties Reduce Workplace Injuries," *Claims Journal*, November 23, 2015, https://www.claimsjournal.com/news/national/2015/11/23/267189.htm.

71. Senate hearing 110–895, "When a Worker Is Killed," 46.

72. Ibid., 4.

73. Ibid., 14.

74. "Occupational Safety and Health Administration (OSHA) Enforcement," U.S. Department of Labor, https://www.osha.gov/dep/2013_bnforcement_summary.html.

75. "Revisions to the 2014 Census of Fatal Occupational Injuries (CFOI)," website of the Bureau of Labor Statistics, U.S. Department of Labor, last modified April 21, 2016, http://www.bls.gov/iif/cfoi_revised14.htm.

76. Christopher P. Lu and Sally Quillian Yates, "Memorandum of Understanding between the U.S. Departments of Labor and Justice on Criminal Prosecutions of Worker Safety Laws," December 17, 2015, https://www.justice.gov/enrd/file/800526/download.

77. Ibid.

78. 33 U.S. Code § 1319(c)(1)(A), "Enforcement," https://www.law.cornell.edu/uscode/text/33/1319.

79. Ibid., (c)(2)(B).

80. Ibid., (c)(3)(A).

81. "Criminal Provisions of the Clean Water Act," U.S. Environmental Protection Agency, https://www.epa.gov/enforcement/criminal-provisions-clean-water-act.

82. "Fiscal Year 2013 EPA Enforcement and Compliance Annual Results," Office of Enforcement and Compliance Insurance, U.S. Environmental Protection Agency, January 13, 2013, http://www.law.uh.edu/faculty/thester/courses/Environmental-Enforcement-2014/fy-2013-enforcement-annual-results-charts-2-6-14_0.pdf.

83. Carey L. Biron, "Criminal Prosecution Rates for Corporate Environmental Crimes Near Zero," *Mint Press News*, July 25, 2014, http://www.mintpressnews.com/criminal-prosecution-rates-for-corporate-environmental-crimes-near-zero/194479/.

84. Brittny Mejia and Veronica Roja, "Bumble Bee Foods to Pay $6 Million in Death of Worker in Pressure Cooker," *Los Angeles Times*, August 12, 2015, http://www.latimes.com/local/lanow/la-me-ln-bumble-bee-worker-killed-settlement-20150812-story.html.

85. "Roofing Company Owner Sentenced for Charges Connected to Employee's Fatal Fall," press release, United States Attorneys Office, Eastern District of Pennsylvania, U.S. Department of Justice, https://www.justice.gov/usao-edpa/pr/roofing-company-owner-sentenced-charges-connected-employees-fatal-fall.

86. Robert Salonga, "Milpitas: Owner, Project Manager Convicted of Manslaughter in 2012 Construction Cave-In," *Mercury News*, May 29, 2015, http://www.mercurynews.com/2015/05/19/milpitas-owner-project-manager-convicted-of-manslaughter-in-2012-construction-cave-in/.

87. Jason Green, "San Francisco Business Owner Charged with Manslaughter after Two Employees Die on the Job," *Mercury News*, July 14, 2016, http://www.mercurynews.com/2016/07/14/san-francisco-business-owner-charged-with-manslaughter-after-two-employees-die-on-the-job/.

88. J. Davitt McAteer et al., *Upper Big Branch: The April 5, 2010, Explosion; A Failure of Basic Coal Mine Safety Practices*, Report to the Governor by the Governor's Independent Investigation Panel, May 2011, http://www.npr.org/documents/2011/may/giip-massey-report.pdf.

89. Adam Liptak, "Motion Ties W. Virginia Justice to Coal Executive," *New York Times*, January 15, 2008, http://www.nytimes.com/2008/01/15/us/15court.html.

90. Capterton v. A.T. Massey Coal Co., Inc., 556 U.S. 868, 884 (2009), https://supreme.justia.com/cases/federal/us/556/868/.

91. "Timeline of Upper Big Branch Mine Disaster Events," *Charleston Gazette-Mail*, November 13, 2014, http://www.wvgazettemail.com/article/20141113/GZ01/141119618.

92. Ibid.

93. Jared Hunt, "Ex-Massey CEO Don Blankenship Indicted by Federal Grand Jury," *Charleston Gazette-Mail*, November 13, 2014, http://www.wvgazettemail.com/article/2014 1113/DM01/141119628.

94. Indictment at 34–43, United States v. Blankenship, 5:14-cr-00244 (S.D.W. Va., Nov. 13, 2014).

95. Ibid., 9.

96. Ibid., 16.

97. Ibid.,18–19.

98. Ibid., 18.

99. Ibid., 19.

100. Ibid., 27.

101. Ibid., 30.

102. Ibid., 23–24.

103. Hunt, "Ex-Massey CEO Don Blankenship Indicted."

104. Ibid.

105. Kirsten Korosec, "Massey CEO Don Blankenship's Retirement Package: $12M in Cash, Health Insurance, a Secretary and a Chevy," CBS Money Watch, last updated December 8, 2010, http://www.cbsnews.com/news/massey-ceo-don-blankenships-retirement-package-12m-in-cash-health-insurance-a-secretary-and-a-chevy/.

106. Hunt, "Ex-Massey CEO Don Blankenship Indicted."

107. Mandi Cardosi, "Update: Gag Order Issued in Don Blankenship, Former Massey Energy CEO, Case," WOWK TV, December 13, 2014, http://www.tristateupdate.com/story/27379225/don-blakenship-former-massey-energy-ceo-indicated-in-federal-grand-jury.

108. Superseding Indictment at 1, United States v. Blankenship, 5:14-cr-00244 (S.D.W. Va., Mar. 10, 2015).

109. Motion to Dismiss at 1, United States v. Blankenship, 5:14-cr-00244 (S.D.W. Va., Feb. 6, 2015).

110. Ken Ward Jr., "In Unsealed Filings, Blankenship Claims 'Vindictive' Prosecution," *Charleston Gazette-Mail*, March 5, 2015, http://www.wvgazettemail.com/article/20150305/GZ01/150309505.

111. Ibid.

112. Associated Press, "Ex-Coal CEO Don Blankenship's Trial Delayed to October," *Washington Times*, June 13, 2015, http://www.washingtontimes.com/news/2015/jun/13/ex-coal-ceo-don-blankenships-trial-delayed-to-octo/.

113. "Black Lung Benefits Act Rule Proposal Protects Coal Miners' Health," news release, U.S. Department of Labor, April 28, 2015, http://www.dol.gov/newsroom/releases/owcp/owcp20150812.

114. Bill Estep and John Cheves, "After Decades of Decline, Black Lung on the Rise in Eastern Kentucky," *Lexington Herald Leader*, July 6, 2013, http://www.kentucky.com/news/special-reports/fifty-years-of-night/article44432793.html.

115. "Mining Fatalities Increased in 2014," *Industrial Safety and Hygiene News*, April 17, 2015, http://www.ishn.com/articles/101185-mining-fatilities-increased-in-2014.

116. Joel Ebert and Ken Ward Jr., "Friday Update: What Blankenship Jurors Were Asked," *Charleston Gazette-Mail*, October 2, 2015, http://www.wvgazettemail.com/blankenship-trial/20151002/friday-update-what-blankenship-jurors-were-asked.

117. Ken Ward Jr., "Blankenship Trial Examined WV's Complex Ties to Coal," *Charleston Gazette-Mail*, December 5, 2015, http://www.wvgazettemail.com/blankenship-trial/20151205/blankenship-trial-examined-wvs-complex-ties-to-coal.

118. Ibid.

119. Ibid.

120. Ken Ward Jr., "Jury Didn't Know Safety Conspiracy Carried Just 1 Year in Prison," *Charleston Gazette-Mail*, December 12, 2015, http://www.wvgazettemail.com/blankenship-trial/20151212/jury-didnt-know-mine-safety-conspiracy-carried-just-1-year-in-prison.

121. Ken Ward Jr., "Experts Predict Wide Targeting during Blankenship Appeal," *Charleston Gazette-Mail*, December 4, 2015.

122. Ward, "Jury Didn't Know Safety Conspiracy."

123. Joel Ebert, "Blankenship Juror: 'We Did the Right Thing,' " *Charleston Gazette-Mail*, December 6, 2015, http://www.wvgazettemail.com/blankenship-trial/20151206/blankenship-juror-we-did-the-right-thing-video.

124. "The Dirty Work of a Coal Baron Exposed," editorial, *New York Times*, December 8, 2015, http://www.nytimes.com/2015/12/08/opinion/the-dirty-work-of-a-coal-baron-exposed.html.

125. "The Blankenship Verdict," opinion, *Wall Street Journal*, December 6, 2015, http://www.wsj.com/articles/the-blankenship-verdict-1449446531.

126. "Fitting Penalties for Mine Safety Violations," editorial, *Charleston Gazette-Mail*, December 22, 2015, http://www.wvgazettemail.com/gazette-opinion/20151222/gazette-editorial-fitting-penalties-for-mine-safety-violations.

127. "H.R. 1926–114th Congress: Robert C. Byrd Mine Safety Protection Act of 2015," November 17, 2016, https://www.govtrack.us/congress/bills/114/hr1926.

128. Ibid.

129. Kate White, "Federal Magistrate Lowers Blankenship Bond, Allows Travel within U.S.," *Charleston Gazette-Mail*, February 27, 2016, http://www.wvgazettemail.com/blankenship-trial/20151228/federal-magistrate-lowers-blankenship-bond-allows-travel-within-us.

130. Ken Ward Jr., "Ex-Massey CEO Blankenship Now in California Federal Prison," *Charleston Gazette-Mail*, May 12, 2016, http://www.wvgazettemail.com/news/20160512/ex-massey-ceo-blankenship-now-in-california-federal-prison.

131. Paul O'Leary et al., "Workplace Injuries and the Take-Up of Social Security Disability Benefits," *Social Security Bulletin* 72, no. 3 (2012), https://www.ssa.gov/policy/docs/ssb/v72n3/v72n3p1.html.

132. Ibid.

133. Gregory P. Guyton, "A Brief History of Workers' Compensation," *Iowa Orthopaedic Journal* 19 (1999): 106–10, https://www.ncbi.nlm.nih.gov/pmc/articles/PMC1888620/.

134. Ibid.

135. Ibid.

136. Ibid.

137. David F. Utterback, Alysha R. Meyers, and Steven J. Wurzelbacher, *Workers' Compensation Insurance: A Primer for Public Health*, Department of Health and Human Services,

Centers for Disease Control and Prevention, National Institute for Occupational Safety and Health, January 2014, https://www.cdc.gov/niosh/docs/2014-110/pdfs/2014-110.pdf.

138. Ibid.

139. *Adding Inequality to Injury: The Costs of Failing to Protect Workers on the Job*, report, U.S. Department of Labor, Occupational Safety and Health Administration, March 2015, https://www.dol.gov/osha/report/20150304-inequality.pdf.

140. Michael Grabell and Howard Berkes, "The Demolition of Workers' Comp," *ProPublica*, March 4, 2015, https://www.propublica.org/article/the-demolition-of-workers-compensation.

141. Ibid.

142. *Adding Inequality to Injury*.

143. Ibid.

144. Utterback, Meyers, and Wurzelbacher, *Workers' Compensation Insurance*.

145. Mandy Locke and Franco Ordonez, "Taxpayers and Workers Gouged by Labor Law Dodge," *Charlotte Observer*, September 4, 2014, http://www.charlotteobserver.com/news/business/article9160493.html.

146. *Adding Inequality to Injury*.

147. Molly Redden, "Walmart, Lowe's, Safeway, and Nordstrom Are Bankrolling a Nationwide Campaign to Gut Workers' Comp," *Mother Jones*, March 26, 2015, http://www.motherjones.com/politics/2015/03/arawc-walmart-campaign-against-workers-compensation.

148. Grabell and Berkes, "Demolition of Workers' Comp."

149. "Injured, Ill, and Silenced: Systemic Retaliation and Coercion by Employers against Injured Workers," National Economic and Social Rights Initiative, April 2015, http://www.nesri.org/resources/injured-ill-and-silenced-systemic-retaliation-and-coercion-by-employers-against-injured-workers.

150. Theodore Roosevelt, "Message to Congress on Workers' Compensation," January 31, 1908, archived on the American Presidency Project website, www.presidency.ucsb.edu/ws/?pid=69649.

151. American Medical Association, *Code of Medical Ethics*, Opinion 10.03: Patient-Physician Relationship in the Context of Work-Related and Independent Medical Examinations.

152. Grabell and Berkes, "Demolition of Workers' Comp."

153. 29 U.S. Code § 654, 5(a)1, "Duties of Employers and Employees," https://www.law.cornell.edu/uscode/text/29/654.

154. 5 U.S. Code §§ 551–559, subchapter 2, "Administrative Procedure," https://www.law.cornell.edu/uscode/text/5/part-I/chapter-5/subchapter-II.

155. Steven P. Croley, *Regulation and Public Interests: The Possibility of Good Regulatory Government* (Princeton, NJ: Princeton University Press, 2008), 14.

156. "A Guide to the Rulemaking Process," Office of the Federal Register, https://www.federalregister.gov/uploads/2011/01/the_rulemaking_process.pdf.

157. Ronald Reagan, "Inaugural Address," January 20, 1981, archived on the American Presidency Project website, www.presidency.ucsb.edu/ws/?pid=43130.

158. Croley, *Regulation and Public Interests*, 15.

159. Jeremy W. Peters, "For Tax Pledge and Author, a Test of Time," *New York Times*, November 19, 2012.

160. Croley, *Regulation and Public Interests*, 15.

161. Ibid., 17.

162. Pedro Nicolaci da Costa, "Bernanke: 2008 Meltdown Was Worse Than Great Depression," *Real Time Economics* blog, *Wall Street Journal*, August 26, 2014, http://blogs.wsj.com/economics/2014/08/26/2008-meltdown-was-worse-than-great-depression-bernanke-says/.

163. "Wall Street Reform: The Dodd-Frank Act," last accessed November 17, 2016, https://www.whitehouse.gov/economy/middle-class/dodd-frank-wall-street-reform.

164. "Brief Summary of the Dodd-Frank Wall Street Reform and Consumer Protection Act," democrats.financialservices.house.gov/uploadedfiles/4173briefsummaryofd-f.pdf.

165. Gary Rivlin, "How Wall Street Defanged Dodd-Frank," *Nation*, April 30, 2013, https://www.thenation.com/article/how-wall-street-defanged-dodd-frank.

166. Jonathan Weisman and Eric Lipton, "In New Congress, Wall St. Pushes to Undermine Dodd-Frank Reform," *New York Times*, January 13, 2015, http://www.nytimes.com/2015/01/14/business/economy/in-new-congress-wall-st-pushes-to-undermine-dodd-frank-reform.html.

167. "Dodd-Frank Progress Report," Davis Polk law firm, https://www.davispolk.com/Dodd-Frank-Rulemaking-Progress-Report.

168. "This Way to Jobs," U.S. Chamber of Commerce, https://www.uschamber.com/sites/default/files/legacy/ads/16011_Politico_Regulatory_Ad_FIN.pdf.

169. Citizens United v. Federal Election Commission, 558 U.S. 310 (2010), https://www.supremecourt.gov/opinions/09pdf/08-205.pdf.

170. "Silica, Crystalline," Occupational Safety and Health Administration, U.S. Department of Labor, https://www.osha.gov/dsg/topics/silicacrystalline.

171. "Health Effects of Occupational Exposure to Respirable Crystalline Silica," National Institute for Occupational Safety and Health Publication No. 2002-129, April 2002, https://www.cdc.gov/niosh/docs/2002-129/.

172. Ki Moon Bang et al., "Silicosis Mortality Trends and New Exposures to Respirable Crystalline Silica—United States, 2001–2010," Morbidity and Mortality Weekly Report, Centers for Disease Control and Prevention, February 13, 2015, https://www.cdc.gov/mmwr/preview/mmwrhtml/mm6405a1.htm.

173. "Silica, Crystalline."

174. "OSHA's Final Rule to Protect Workers from Exposure to Respirable Crystalline Silica," Occupational Safety and Health Administration, U.S. Department of Labor, https://www.osha.gov/silica/.

175. Hoeven Silica Amendment, Senate Appropriations Committee, http://www.appropriations.senate.gov/imo/media/doc/hearings/FY16-LaborHHS-Hoeven-Amendment-Silica.pdf.

176. Sen. John Hoeven Summary Data, Center for Responsive Politics, https://www.opensecrets.org/politicians/summary.php?cid=N00031688.

177. "North Dakota Energy Facts," North Dakota Energy Forum, http://www.ndenergyforum.com/expert-facts/.

178. "Worker Exposure to Silica during Hydraulic Fracturing," hazard alert, Occupational Safety and Health Administration, U.S. Department of Labor, https://www.osha.gov/dts/hazardalerts/hydraulic_frac_hazard_alert.html.

179. "U.S. Department of Labor Announces Final Rule to Improve U.S. Workers' Protection from the Dangers of Respirable Silica Dust," news release, U.S. Department of Labor, March 24, 2016, https://content.govdelivery.com/accounts/USDOL/bulletins/13ef9fc.

180. Lydia Wheeler, "Industry Groups Ask Federal Court to Review DOL's Silica Rule," *Hill*, April 4, 2016, http://thehill.com/regulation/court-battles/275110-industry-groups-ask-federal-court-to-review-dols-silica-rule.

181. "OSHA's Ergonomics Regulation," U.S. Chamber of Commerce, August 4, 2010, https://www.uschamber.com/oshas-ergonomics-regulation.

182. "Ergonomics Background," Occupational Safety and Health Administration, U.S. Department of Labor, https://www.osha.gov/archive/ergonomics-standard/archive.html.

183. "Ergonomics," Safety and Health Topics, Occupational Safety and Health Administration, U.S. Department of Labor, https://www.osha.gov/SLTC/ergonomics/.

184. "Cranes and Derricks in Construction, Final Rule," *Federal Register* 75, No. 152, August 9, 2010, Occupational Safety and Health Administration, U.S. Department of Labor, https://www.osha.gov/pls/oshaweb/owadisp.show_document?p_table=FEDERAL_REGISTER&P_id=21692.

185. Ibid.

186. "Delays in Recent OSHA Safety and Health Standards Impact on Workers' Lives" (table), AFL-CIO, www.aflcio.org/content/download/126581/3464331/27+Delays+in+OSHA +Standards+Impact+On+Workers+2014.pdf.

187. "Cranes and Derricks in Construction, Final Rule."

188. Ashley Southall, "Crane Accident in Midtown Manhattan Injures 7 and Damages Buildings," *New York Times*, May 31, 2015, http://www.nytimes.com/2015/06/01/nyregion/crane-collapses-in-midtown-manhattan.html.

189. "Beryllium," Safety and Health Topics, Occupational Safety and Health Administration, U.S. Department of Labor, https://www.osha-gov/SLTC/beryllium.

190. "Occupational Exposure to Beryllium and Beryllium Compounds; Proposed Rule," *Federal Register* 80, No. 152, August 7, 2015, Occupational Safety and Health Administration, U.S. Department of Labor, https://www.osha.gov/pls/oshaweb/owadisp. show_document?p_table=FEDERAL_REGISTER&p_id=25346.

191. "Hazard Communication Guidance for Combustible Dusts," OSHA 3371-08, 2009, Occupational Safety and Health Administration, U.S. Department of Labor, https:// www.osha.gov/Publications/3371combustible-dust.html.

192. Ibid.

193. "Worker Protection against Combustible Dust Explosions and Fires Act of 2008, H.R. 5522, 110th Congress," on GovTrack.us, https://www.govtrack.us/congress/bills/110/hr5522.

194. "Rulemaking Watch: OSHA's Combustible Dust Rulemaking," U.S. Chemical Safety Board, last updated May 26, 2016, www.csb.gov/mobile/mw/dust/regwatch/.

195. Ibid.

196. Brian Dabbs, "OSHA Poised to Pass on Combustible Dust in 2016," *Bloomberg News*, January 14, 2016, http://www.bna.com/osha-poised-pass-n57982066224/.

197. Timeline on Combustible Dust Activities, www.csb.gov/recommendations/timeline/.

198. Bruce Rolfsen, "New OSHA Injury Reporting Standard Challenged in Court," *Bloomberg BNA*, July 10, 2016, http://www.bna.com/new-osha-injury-n73014444007/.

199. *Federal Register* 80, No. 145, July 29, 2015, Occupational Safety and Health Administration, U.S. Department of Labor, https://www.osha.gov/FedReg_osha_pdf/FED20150729A.pdf.

200. A group of employers and business associations, led by the U.S. Chamber of Commerce, and calling itself the "Coalition for Workplace Safety," opposed the *Volks* rule clarification. In a letter to OSHA on October 27, 2015, the CWS submitted a twelve-page comment demanding that OSHA withdraw the proposed regulation. Coalition for Workplace Safety, letter to David Michaels (assistant secretary for occupational safety and health), October 27, 2015, http://www.ipc.org/3.0_Industry/3.4_EHS/2015/CWS-Comments-on-Volks-NPRM.pdf.

201. Phillip Rucker and Robert Costa, "Bannon Vows a Daily Fight for 'Deconstruction of the Administrative State,'" *Washington Post*, February 23, 2017, https://www. washingtonpost.com/politics/top-wh-strategist-vows-a-daily-fight-for-deconstruction-of-the-administrative-state/2017/02/23/03f6b8da-f9ea-11e6-bf01-d47f8cf9b643_story.html?utm_ term=.0f9a1aa7d586.

202. Department of Defense (DoD), General Services Administration (GSA), and National Aeronautics and Space Administration (NASA), *Federal Register*, "Federal Acquisition Regulation: Fair Pay and Safe Workplaces," final rule, August 25, 2016, https://www.federal register.gov/documents/2016/08/25/2016-19676/federal-acquisition-regulation-fair-pay-and-safe-workplaces.

203. Debbie Berkowitz, "Who's Against Fair Pay and Safe Workplaces?," *The Hill*, March 7, 2017, http://thehill.com/blogs/congress-blog/labor/322711-whos-against-fair-pay-and-safe-workplaces.

204. Croley, *Regulation and Public Interests*, 99–101.

205. Ibid., 101.

206. Brian Wolfman and Bradley Girard, "The Court Slays the D.C. Circuit's *Paralyzed Veterans* Doctrine, Leaving Bigger Issues for Another Day," *SCOTUS blog*, March 10, 2015, http://www.scotusblog.com/2015/03/opinion-analysis-the-court-slays-the-d-c-circuits-para lyzed-veterans-doctrine-leaving-bigger-issues-for-another-day/.

207. White House Safeguard Tracker, a Project of Public Citizen, www.safeguardsde layed.org.

208. Ibid.

209. GovTrack website, S.Con.Res. 17 (114th): RESTORE Resolution of 2015, https://www.govtrack.us/congress/bills/114/sconres17/details.

210. Ibid.

211. "Rounds Introduces RESTORE to Permanently Address Regulatory Reform," press release, Mike Rounds senate website, May 20, 2015, www.rounds.senate.gov/newsroom/press-releases/rounds-introduces-restore-to-permanently-address-regulatory-reform.

212. Denise Robbins, "Experts Debunk *WSJ*'s Favorite Report on Cost of Federal Regulations," Media Matters for America, May 18, 2015, http://mediamatters.org/research/2015/05/18/experts-debunk-wsjs-favorite-report-on-cost-of/203684.

213. Juliet Eilperin, "Anatomy of a Washington Dinner: Who Funds the Competitive Enterprise Institute?," *Washington Post*, June 20, 2013, https://www.washingtonpost.com/news/the-fix/wp/2013/06/20/anatomy-of-a-washington-dinner-who-funds-the-competitive-enterprise-institute/.

214. Office of the President of the United States, Office of Management and Budget, Office of Information and Regulatory Affairs, "2016 Draft Report to Congress on the Benefits and Costs of Federal Regulations and Agency Compliance with the Unfunded Mandates Reform Act," https://obamawhitehouse.archives.gov/sites/default/files/omb/assets/legislative_re ports/draft_2016_cost_benefit_report_12_14_2016_2.pdf.

215. The White House, President Donald J. Trump, "Implementation of Regulatory Freeze," M-17-16, memorandum, January 20, 2017, https://www.whitehouse.gov/the-press-office/2017/01/24/implementation-regulatory-freeze.

216. The White House, President Donald J. Trump, Office of the Press Secretary, "Presidential Executive Order on Reducing Regulation and Controlling Regulatory Costs," press release, January 30, 2017, https://www.whitehouse.gov/the-press-office/2017/01/30/presidential-executive-order-reducing-regulation-and-controlling.

217. Lydia Wheeler and Lisa Hagen, "Trump Signs '2- for -1' Order to Reduce Regulations," *The Hill*, January 30, 2017, http://thehill.com/homenews/administration/316839-trump-to-sign-order-reducing-regulations.

218. Karla Walter, David Madland, Paul Sonn, and Tsedeye Gebreselassie, "Contracting That Works," Center for American Progress Action Fund, November 13, 2015, https://www.americanprogressaction.org/issues/economy/reports/2015/11/13/125359/contracting-that-works-2/.

219. Associated Builders and Contractors, "Responsible Contractor Ordinances," http://www.mansfieldct.gov/filestorage/1904/5335/1912/21835/20121018_info_resp_contracting.pdf.

220. Md. Stat. Ch. 189.

221. "Responsible Contractor Requirement Defined," Minnesota Statute § 16C.285, https://www.revisor.mn.gov/statutes/?id=16C.285.

222. Ibid.

223. Walter et al., "Contracting That Works."

224. Ibid.

225. Santa Clara Valley Transportation Authority, http://www.vta.org/about-us/procure ment/vta-cam.

226. "Fair Pay and Safe Workplaces, Executive Order 113,673, July 31, 2014, https://www.whitehouse.gov/the-press-office/2014/07/31/executive-order-fair-pay-and-safe-workplaces.

227. Ben Penn, "Lobbyists Prepare for Battle over Contractor Fair Pay Rule," *Bloomberg BNA*, March 4, 2016, http://www.bna.com/lobbyists-plan-battle-n57982068114/.

228. "Executive Order 13673: Fair Pay and Safe Workplaces," Office of the Assistant Secretary for Policy, U.S. Department of Labor, htttps://www.dol.gov/asp/fairpayand safeworkplaces/.

7. Are There Really Any Accidents?

1. Worker Fatalities Reported to Federal and State OSHA, U.S. Department of Labor, Occupational Safety and Health Administration, https://www.osha.gov/dep/fatcat/dep_fat cat.html.

2. Ibid.

3. "OSHA Says Worker Safety Pays, Amputations Cost," press release, U.S. Department of Labor, Occupational Safety and Health Administration, September 10, 2015, https://www.osha.gov/pls/oshaweb/owadisp.show_document?p_table=NEWS_RELEASES&p_id=28686.

4. Celeste Monforton, "Amputations Abound at Tyson Foods, OSHA Records Shed More Light on Industrial Food Production," *Pump Handle* blog, Science Blogs, January 27, 2016, http://scienceblogs.com/thepumphandle/2016/01/27.

5. "Water, Rest, Shade: The Work Can't Get Done without Them," U.S. Department of Labor, Occupational Safety and Health Administration, https://www.osha.gov/SLTC/heatill ness/index.html.s.

6. Austin v. Kroger Texas, L.P., No. 14-0126, June 12, 2015, Supreme Court of Texas.

7. "Worked to Death," editorial, *Houston Chronicle*, July 24, 2015, http://www.chron.com/opinion/editorials/article/Worked-to-death-6404426.php.

8. Diana Florence, phone interview with the author, December 29, 2016.

INDEX

Occupational Safety and Health
Administration (OSHA) *(continued)*
silica exposure safety and, 194–195
Volks rule, 199–200, 237n200
voluntary guidelines, reliance on, 42,
52, 62, 66, 95–96, 143, 196
warehouse safety and, 119–120
weakening/limiting of, 34–35, 37–38,
40–45, 88, 109, 156–157, 195, 199
oil and gas workers, 4, 40, 44, 45,
85–93, 195, 211
OIRA (Office of Information and
Regulatory Affairs), 201–202
Oklahoma, 183
Oppegard, Tony, 82–83
organized labor. *See* labor unions

packinghouse workers, 121–129
PAWA (Protecting America's Workers
Act), 162–163, 166
penalties
average/median, 41, 163–164, 165
effectiveness of, 167
environmental penalties *vs.*, 164,
169–170
legal maximums, 164–165, 167
specific cases: barge death, 100; boiler
explosion death, 92; construction
worker deaths, 44, 170–171;
electrocution death, 76; forklift
operator deaths, 58–59, 60,
119–120; grain bin death, 140, 141;
serial offender, 166; sulfuric acid
death, 164
systematic discounting of, 165–166
Pennsylvania, 170–171
Perez, Tom, 193
Perkins, Frances, 23, 24, 26, 32
personal protective equipment, 68,
75–76
PhilaPOSH, 100
pipe fitters, 85–93
Pittsburgh Survey, 28–30
political space, 163
poultry plant workers, 121–129

presidential vetoes, 201–202
public choice theory of regulation, 191
Public Citizen, 201–202
publicly financed projects, 206–208
public opinion, 6, 15–16, 17, 23, 24, 28

railroad workers, 12, 30
RCOs (responsible contracting
ordinances), 206–207
Reagan, Ronald, 190, 191
regulation and legislation
bills/orders to prevent, 202–204, 205
coal mine safety, 30–31, 78, 79, 177
delaying tactics to prevent, 163,
192–198, 200–202
electrical safety, 68, 69
federal, 30–38, 78, 79, 185–205
health sector safety, 142, 143
international comparisons, 12
lobbying against, 3, 25–26, 32–38,
80, 123–124, 154–161, 183–184,
190–193; by U.S. Chamber Of
Commerce, 26, 33, 50, 192, 196,
200, 207, 237n200
oil/gas industry safety, 88
race to the bottom, 28, 161, 185–186
reforms needed: compensation system,
184–187; regulatory system,
187–205
repeal of, 200, 202–203
Republicans' weakening of, 34–35,
163, 177, 194–196, 200, 202–205
state, 23–28, 129, 143, 160–161, 173,
180–184, 205–207
See also OSHA
Regulation journal, 155–161
regulatory agencies, role of,
188–190, 191
repetitive strain disorders, 49–50, 61–67,
122–129
Republicans, hostility to safety
regulation, 34–35, 163, 177,
194–196, 200, 202–205
RESTORE bill, 202–204
retail workers, 47–60